D0082789

VICTORIAN TECHNOLOGY

VICTORIAN TECHNOLOGY

Invention, Innovation, and the Rise of the Machine

Herbert Sussman

VICTORIAN LIFE AND TIMES
Sally Mitchell, Series Editor

PRAEGER
An Imprint of ABC-CLIO, LLC

A B C 📖 C L I O

Santa Barbara, California • Denver, Colorado • Oxford, England

Library of Congress Cataloging-in-Publication Data
Sussman, Herbert L.
   Victorian technology : invention, innovation, and the rise of the machine / Herbert Sussman.
      p. cm. — (Victorian life and times)
   Includes bibliographical references and index.
   ISBN 978-0-275-99169-2 (hardcopy : alk. paper) — ISBN 978-0-313-08285-6 (ebook) 1. Technology—Great Britain—History—19th century. 2. Machinery—Great Britain—History—19th century. 3. Inventions—Great Britain—History—19th century. 4. Technological innovations—Great Britain—History—19th century. I. Title.
   T26.G7S95 2009
   609.41′09034—dc22   2009018904

13   12   11   10   9   1   2   3   4   5

This book is also available on the World Wide Web as an eBook.
Visit www.abc-clio.com for details.

ABC-CLIO, LLC
130 Cremona Drive, P.O. Box 1911
Santa Barbara, California 93116-1911

This book is printed on acid-free paper ∞

Manufactured in the United States of America

TO
HENRY, OWEN, JACOB, QUILL

CONTENTS

SERIES FOREWORD

Although the nineteenth century has almost faded from living memory—most people who heard firsthand stories from grandparents who grew up before 1900 have adult grandchildren by now—impressions of the Victorian world continue to influence both popular culture and public debates. These impressions may well be vivid yet contradictory. Many people, for example, believe that Victorian society was safe, family-centered, and stable because women could not work outside the home, although every census taken during the period records hundreds of thousands of female laborers in fields, factories, shops, and schools as well as more than a million domestic servants—often girls of fourteen or fifteen—whose long and unregulated workdays created the comfortable leisured world we see in Merchant and Ivory films. Yet it is also true that there were women who had no household duties and desperately wished for some purpose in life but found that social expectations and family pressure absolutely prohibited their presence in the workplace.

The goal of books in the Victorian Life and Times series is to explain and enrich the simple pictures that show only a partial truth. Although the Victorian period in Great Britain is often portrayed as peaceful, comfortable, and traditional, it was actually a time of truly breathtaking change. In 1837, when eighteen-year-old Victoria became queen, relatively few of England's people had ever traveled more than ten miles from the place where they were born. Little more than half the population could read and write, children as young as five worked in factories and mines, and political power was entirely in the hands of a small minority of men who held property. By the time Queen Victoria died in 1901, railways provided fast and cheap transportation for both goods and people, telegraph messages sped to the far corners of the British Empire in minutes, education was compulsory, a man's religion (or lack of it) no longer barred him from sitting in Parliament, and women were not only wives and domestic servants but also physicians, dentists, elected school-board members, telephone operators, and university lecturers. Virtually every aspect of life had been transformed either by technology or by the massive political and legal reforms that reshaped Parliament, elections, universities, the army, education, sanitation, public health, marriage, working conditions, trade unions, and civil and criminal law.

The continuing popularity of Victoriana among decorators and collectors, the strong market for historical novels and for mysteries set in the age of Jack the Ripper and Sherlock Holmes, the new interest in books by George Eliot and Charles Dickens and Wilkie Collins whenever one is presented on television, and the desire of amateur genealogists to discover the lives, as well as the names, of nineteenth-century British ancestors all reveal the need for accurate information about the period's social history and material culture. In the years since my book, *Daily Life in Victorian England*, was published in 1996, I have been contacted by many people who want more detailed information about some area covered in that overview. Each book in the Victorian Life and Times series will focus on a single topic, describe changes during the period, and consider the differences between country and city, between industrial life and rural life, and above all, the differences made by class, social position, religion, tradition, gender and economics. Each book is an original work, illustrated with drawings and pictures taken from Victorian sources, enriched by quotations from Victorian publications, based on current research, and written by a qualified scholar. All of the authors have doctoral degrees and many years' experience in teaching; they have been chosen not only for their academic qualifications but also for their ability to write clearly and to explain complex ideas to people who do not have extensive background in the subject. Thus, the books are authoritative and dependable but written in straightforward language; explanations are supplied whenever specialized terminology is used; and a bibliography lists resources for further information.

The Internet has made it possible for people who cannot visit archives and reference libraries to conduct serious family and historical research. Careful hobbyists and scholars have scanned large numbers of primary sources—nineteenth-century cookbooks, advice manuals, maps, city directories, magazines, sermons, church records, illustrated newspapers, guidebooks, political cartoons, photographs, paintings, published investigations of slum conditions and poor people's budgets, political essays, inventories of scientists' correspondence, and many other materials formerly accessible only to academic historians. Yet the World Wide Web also contains misleading documents and false information, even on educational sites, created by students and enthusiasts who don't have the experience to put material in useful contexts. So far as possible, therefore, the bibliographies for books in the Victorian Life and Times series will also offer guidance on using publicly available electronic resources.

Technology was responsible for the most readily visible changes in nineteenth-century Britain—in many cases literally visible, as in the bridges and railway train sheds that are still in use, and, more generally, in the revolutionary advances in transportation, communication, and industrial production that spring to mind when we think about the differences between life at the century's beginning and at its end. In *Victorian Technology: Invention, Innovation, and the Rise of the Machine*, Herbert Sussman describes the working of new machines and industrial processes as well as the men who invented or developed them. In addition, he explains the ways that technology changed not only the physical world but also the ideologies that governed social life, the organization of national and international institutions, people's relationships with one another, and their beliefs about the world. Many of these alterations are now so familiar to us that we take them as normal and fail to notice them. Considering fields as diverse as law and popular culture, patents and finance, the standardization of time, the conquest of distance, and the regulation of working life, he demonstrates the foundation of still-troubling questions about the relationship of human beings to the machines that have become, over the past two hundred years, so central to human life.

Sally Mitchell, Series Editor

PREFACE

In writing this book, I have tried to present for the general reader an account of Victorian technology in the broadest sense of the term technology. The volume describes not only the development of such new forms of the machine in the nineteenth century as the computer and the Internet, but also the complex ways that technological innovation was connected to the values of the age. In the nineteenth century, as in our own time, the development of technologies was dependent upon the ideas of the society. In turn, the new technologies shaped the culture. Our own machine-dominated time, like that of the Victorians, still engages the problem of how to live a human and humane life following the rise of the machine.

Like the process of invention and innovation, the writing of a book depends on the work of others. In particular I would mention conversations with fellow Victorianists Adrienne Munich, John Maynard, and Carole Silver. I had the opportunity to present material from this study and benefit from helpful comments at the CUNY Victorian Seminar, Northeast Victorian Studies Association, and St. Louis University. Great thanks to my editor Sally Mitchell for her unrivaled expertise and steady guidance. And, of course, to Elisabeth.

CHRONOLOGY

1804   Richard Trevithick builds the first steam locomotive operating on tracks

1807   Robert Fulton sails the *Clermont* on the Hudson River to begin the first commercial steamboat service in the world

1812   Parliament passes Frame-Breaking Act in effort to control the Luddite movement

1822   Charles Babbage completes prototype of the Difference Engine

1825   Stockton and Darlington Railway opens

1830   Liverpool and Manchester Railway opens

1831   Michael Faraday builds the induction ring and Faraday disc to generate electricity

1833   Factory Act passed by Parliament to regulate child labor

1834   Founding of Grand National Consolidated Trades Union

1837   Victoria becomes queen

        First commercial electrical telegraph patented by Sir William Fothergill Cooke

1838   Beginning of scheduled transatlantic steamship service with sailing of *Great Western* from Bristol, England to New York

1839   Chartism petition presented to Parliament. Parliament refuses to accept petition

        First use of railway telegraph, on the Great Western Railway

1840   Charles Babbage announces design of Analytical Engine

        George and Henry Elkington awarded the first patents for electroplating

1843   Completion of tunnel for pedestrians under Thames River, incorporated into London Underground system in 1869

1844   Factory Act to improve conditions of employment for women and children in textile mills

        Joint Stock Companies Act

1847   Ten Hours Act further limiting hours of employment of women and young persons

1848   Parliament rejects third Chartist petition. Chartism movement dissolves

        Public Health Act providing for a Central Board of Health

        Greenwich Mean Time adopted by all railway companies

1850   Britannia Bridge of Robert Stephenson completed

        Anglo-French Telegraph Company lays first undersea telegraph cable, beneath English Channel

1851   Great Exhibition of the Works of Industry of All Nations opens

1852   Undersea telegraph cable between Great Britain and Ireland

1853   George Cayley's glider carries out first recorded flight of a person in an aircraft

1855   Limited Liability Act limiting stockholder financial responsibility

1858   First undersea telegraph cable between England and America completed, but soon fails

1859   Royal Albert railway bridge at Saltash, Cornwall by Isambard Kingdom Brunel completed

1860   British Navy builds HMS *Warrior*, its first steam-powered ironclad

1862   Machine gun invented by an American, Richard Gatling

        First naval battle of steam-powered iron warships, between the *Merrimack* and *Monitor* during American Civil War

1863   Metropolitan Railway, world's first urban underground passenger-carrying railway

1866   Completion of a reliable transatlantic cable

1876   Alexander Graham Bell makes the first telephone call

1878   Joseph W. Swan, an Englishman, granted British patent for incandescent light bulb

        Commercial long-distance telephone calling begins in England

1879   Thomas Alva Edison, an American, granted American patent for incandescent light bulb

1884   Charles Parsons patents an efficient steam turbine

1885   Gottlieb Daimler of Germany exhibits prototype of the modern gasoline automobile engine

1886   Herbert Ackroyd Stuart patents compression-ignition oil engine

        Karl Benz of Germany patents a gasoline-powered automobile

1888   William Lever creates the model village of Port Sunlight, UK

1889   Maxim gun, the first water-cooled machine gun, adopted by the British Army

1890   City & South London line becomes the first electrically-powered underground line

1891   First submarine telephone cable, connecting London and Paris

        Richard Hornsby and Sons builds the first compression-ignition engine

1895   Lanchester Engine Company builds the first four-wheel gasoline-driven automobile in Britain

1896   Guglielmo Marconi, an Italian, granted in England world's first patent for a system of wireless telegraphy

INTRODUCTION

For the people of the nineteenth century, the age of invention began with a single indelible public moment, the opening of the Liverpool and Manchester Railway in 1830. Writing in 1847, the end of the first decade of Queen Victoria's reign, a Victorian workingman recalled:

> The time I first saw the railway uniting Liverpool and Manchester ... stands, as I have said, in memory, like an epoch in my life. I looked upon that most poetical and most practical of the grand achievements of human intellect, until people thought I stood and slept; and, when they heard the dream, they said it was very dreamy, indeed. I should fear to tell the dreams which I have now beside the electric telegraph, and on the railways, and within the regions of the god-like inventors and makers of machinery. There is a time coming when realities shall go beyond any dreams that have yet been told of these things.... of the moral electricity ... to carry the instantaneous message of one feeling, one interest, one object, one hope of success from the lordly end, to the working man's end of the social world.[1]

On that September day in 1830, thousands of spectators, workingmen with cloth caps, middle-class men in black suits, and women in bonnets lined the rails stretching between the manufacturing city of Manchester and the port of Liverpool to cheer a sight never before seen. Along iron rails on iron wheels a contrivance consisting of a large round barrel shape with the word Rocket on the side came chuffing along the rails at the speed of seventeen miles per hour. From the barrel a tubular pipe rose

The Rocket locomotive built by George Stephenson with its coal-carrying tender. The piston rods moved by steam from the boiler move connecting rods that drive the two large front wheels. The Rocket opened the Liverpool & Manchester Railway in 1830 and ran on that line to 1836. [Reprinted with permission of National Railway Museum/Science & Society Picture Library.]

upward, and white water vapor mixed with black soot emerged from the barrel pipe. In a small cabin behind the barrel was a driver, or in nautical terms a pilot, as well as a man rapidly shoveling coal into the barrel. Attached to what would come to be known as the steam locomotive were the carriages which resembled older stagecoaches, themselves on wheeled platforms, each separate, each with side opening doors and windows. Thus opened the Liverpool and Manchester Railway, the world's first intercity passenger railway in which all the trains were operated solely by steam locomotives.

This glorious day for the inventors, for the financial speculators, for the manufacturers of England, and indeed for the movement of the entire world to the age of steam was marred by a tragic event prophetic of the mixed costs and benefits of the rapid rise of the machine in the nineteenth century. Among the British elite gathered for this spectacle was Sir William Huskisson, a popular Member of Parliament for Liverpool, who seized the opportunity of a temporary halt to alight from his carriage on this first train and talk to the Duke of Wellington, then Prime Minister, through the Duke's carriage window on another train. Standing on the right of way, he mis-judged the speed of the approaching Rocket. Indeed, how could he properly judge the speed of a steam passenger locomotive since one had never before been seen in operation? He was run over, becoming the world's first railway passenger fatality.

Huskisson was not killed instantly, however. In ways foretelling the med-ical benefits of improved transportation, the locomotive Northumbrian was detached from the Duke's train and rushed him to the nearby town of Eccles. There he, alas, died, showing the uneven developments of technologies; medical science had developed more slowly than the technology of steam.

THE AGE OF INVENTION

From 1830 through the 1890s, the English of all classes marveled at the achievements of "the god-like inventors and makers of machinery" who had transformed an agricultural Britain into an industrial nation. By the 1840s, the power of steam had supplanted the power of water wheels, of animals, and of human muscles. The domestic workrooms of hand-loom weavers had been replaced by factories with vast internal spaces filled with machines to spin thread and to weave cloth, all powered by the steam engine. Men and women had migrated from rural villages to the new indus-trial cities to work in these factories. And these factory cities, such as Man-chester and Birmingham as well as the metropolis of London, were linked by the steam engine on wheels, the locomotive running on a vast network of iron tracks.

The reign of Victoria, from the late 1830s through the 1890s, saw the full development in England of technologies that created the world that in many ways we still inhabit. Although the genesis of much of this machinery can be traced back to the eighteenth century, in the early nineteenth century, such inventions as the steam engine, the railway, the coke-fired blast fur-nace producing cast iron, and automatic textile machinery merged into the factory system that provided the model for mechanized industry into our own time.

The men associated with the steam engine, the railway locomotive, the ocean-going iron steamship, and the steam hammer were perceived throughout the society as god-like. They became the culture heroes of the age of invention. Guaranteed economic rewards by new financial methods such as limited-liability joint stock companies, the inventor or, in the words of the nineteenth century, the mechanist, flourished. And the age celebrated the new beauty seen in these inventions—the sleek forms of the engines, the power of the locomotive, the dance of the pistons.

But the very social system that encouraged invention grounded in the ideology of laissez faire that rejected government regulation at the same time generated the social and physical horrors of this early industrial revolution. The newcomers to the industrial city occupied overcrowded hovels in a world of coal smoke and polluted rivers. Within the factories, small particles of cotton waste called fluff hung in the air and coated the lungs; small children worked long hours. The bodies of children, women, and men were torn apart in unfenced machinery. Wages were kept at starvation level. Such conditions generated labor unrest, manifested in industrial strikes and an increasingly organized urban working class. By the 1840s, industrial novels by such figures as Charles Dickens and Elizabeth Gaskell emerged to offer a more humane alternative to industrial capitalism. Art and architecture found value in irregularity and hand labor that were seen as the antitheses of factory production.

The new technologies of steam and the railway were formed by practical men such as James Watt, the developer of the steam engine, who innovated by tinkering in the workshop. Even though the Victorians developed the institutions of pure science, the activity of nineteenth-century scientists was often directed to technological applications. The case of Charles Babbage, a Cambridge professor of mathematics, is exemplary in that he turned from the abstractions of mathematics to employing mathematical principles to build the celebrated Victorian proto-computer he called the Difference Engine.

This, then, is a book about invention and innovation, particularly about the mutually reinforcing effects of new technologies in England in the nineteenth century. But any consideration of Victorian technology must first consider the complex meaning of the word technology itself. Often, perhaps too often, the term carries only the sense of newness and of complexity, as we might say in our own time that technology has made communications easier. And yet it is important to see that any material technique for accomplishing a specific task is a technology. The chipping of flints into arrowheads by using stones is a technology devoted to the end of making efficient instruments for killing animals. Thus, we can speak of a Stone Age technology or even of advances in Stone Age technology.

To take an example closer to the nineteenth century, in the European world, the primary technology for moving heavy objects for long distances over land was the wheeled cart pulled by horses or oxen over dirt roads. Similarly, the power technology for the familiar spinning wheel that transformed raw wool into yarn was the foot-power of the woman pushing the treadle in her home. To weave the yarn into clothing, the main technology was the traditional loom, in which threads were drawn through webs by a shuttle moved by hand.

Thus, when we speak of Victorian technology we do not mean the sudden appearance of invention. After all, the wheel was invented, as was the edged flint arrowhead. Rather, we refer to, and this is the subject of this book, invention and innovation in a historically unprecedented profusion that generated a specific array of techniques in nineteenth-century England and that in many ways created our own mechanized world. More specifically, the nineteenth century saw a technological revolution whose transformative principle was the replacement of the muscle power of animals and human beings—the oxen pulling the cart, the woman pushing the treadle, the hand-loom weaver pulling the threads—with the energy generated by the steam engine and later in the century by electricity. The horse is replaced by the iron horse in the locomotive whose wheels turn regularly and tirelessly, powered by the expansive power of steam generated in the boiler. The multiple spindles in the cotton mill making raw cotton or raw wool into threads are turned by the stationary steam engine. The shuttle is drawn through the web of the power loom by a mechanism powered by steam.

This transformation of the world through Victorian technology can be described succinctly as the rise of the machine. But, as with the term technology, the term machine has several meanings. If we consider the term machine as describing a device for multiplying power, then machines have been around for a long time in such forms as a boat with a sail, a block-and-tackle pulley, the inclined ramp used on the Egyptian pyramids, even the horse-drawn wheeled cart and the hand loom. But, just as the term Victorian technology implies a specific and unprecedented set of devices, so the term Victorian machine describes a new kind of machine. First, the nineteenth-century machine was driven not by horse or wind or water, but by the controlled force of steam and later in the century by electricity. Second, the new machines were self-regulating devices that ran without any need of control by human beings.

These machines, steam-powered and self-regulating, did not appear suddenly in finished form, but rather developed over time through multiple innovations. For example, although it is commonly said that James Watt invented the steam engine, it is more accurate to think of the Victorian steam engine as the result of a series of improvements going back to the eighteenth century in building a machine to transform the heat energy of expanding steam into the mechanical energy that moved a piston. Within this long process, as we shall see, Watt did make several crucial improvements. But Watt's innovations can be seen within a long line of innovations in harnessing steam that continues to our own day. Victorian machines did not exist in a fixed form, but evolved rapidly within a culture that supported innovation.

The central elements of the technological revolution of the nineteenth century are clear:

- The shift from the power of human muscle or animals to mechanical power, particularly that of steam and later in the nineteenth century of electricity
- the movement of production from home or small workshop to centralized production in the factory or mill

- development of self-regulating, automatic machines needing no human control once set in operation
- mass production employing the division of labor in which manufacturing is divided into separate, easily repeated functions
- standardization in the interchangeable parts used in manufacturing and in construction, and even in the standardization of time itself.

THE MEANING OF TECHNOLOGY

Victorian technology was indeed revolutionary in its creation of steam-powered, self-regulating machines. We see in our mind's eye the steam locomotive, a cloud of steam rising from its smokestack as it charges across the pastures of England. Or the vast space of a Victorian factory filled with power looms. But in speaking of Victorian technology or the Victorian machine we must look beyond the locomotive and the loom to the larger systems through which the machine is created and in which the machine functions.

Although we think of a machine such as the locomotive or power loom as a freestanding independent object, such a mechanism exists within a system or set of interdependent technologies. To take a contemporary example, consider the automobile. It is only through the habits of language that we generally think of the car as single machine. An individual car is actually an entity composed of many mechanical and electronic technologies that comprise a multitude of technological systems. Thus, we accurately speak not of brakes, but of the braking system. Increasingly, in a continuation of the Victorian invention of automatic control systems for machines, the braking system and the engine system are regulated not by the driver, but by computers.

Just as what we see as the individual machine is in actuality a set of interdependent technologies, in the nineteenth century as in our own time the machine functions within a large and complex set of economic, social, and cultural systems. As Leo Marx, the distinguished student of nineteenth-century innovation, states, *"technology* narrowly conceived as a physical device ... is merely one part of a complex social and institutional matrix".[2]

As an example of technology seen not merely in terms of individual machines, but as a complex interdependent organization of machines and culture, we can look again to the automobile in America. The use of the car depends upon other machines, such as the robots that manufacture cars; the machines that build the freeways and bridges; the system of traffic lights that regulate the stream of cars; oil-extracting drills, refineries, and pumping stations that provide gasoline; and so on. Furthermore, the efficiency of the car as transport depends upon certain social choices. We commonly speak of America as a car culture, to point to decisions beyond mechanical and electronic innovation. The choice between developing engines that burn less fuel or engines that provide high speed, between producing cars that are smaller to lessen congestion or SUVs to transport families safely are all decisions made by the culture. As a society we have allocated resources for the construction of roads and freeways rather than public transportation, made efforts to keep the price of gasoline affordable, and built suburbs whose inhabitants must depend on cars.

In the Victorian period, the spectacular spread of the railway during the 1830s and 1840s was due to advances in the technology of the steam locomotive. But the railway must be seen as the railway system. The trains ran on iron tracks forged in the new coke-fired blast furnaces. The electric telegraph provided the virtually instantaneous communication needed to regulate a signal system for safe operation. A workforce had to be trained. Society had to choose a standard track width. A system of joint stock companies with limited liability had to be developed to finance the new enterprise. Then, too, it was the widely held belief that the destruction of landscape and cityscape was unavoidable that led to the social decision, similar to that of American car culture, to tear through city neighborhoods for railway tunnels, bridges, and termini.

Thus, not only the individual mechanisms and submechanisms, but also the larger system of social and cultural choices comprise a mechanized society. The Victorians themselves used the word machinery in this extended sense. Describing the ways that new devices were becoming embedded in a new society in daily life and even in individual consciousness, Thomas Carlyle, an influential early-Victorian social critic, in an essay aptly titled "Signs of the Times," called it "the Age of Machinery, in every outward and inward sense of that word."[3]

Separating the machine from its social context carries potentially dangerous implications, particularly in leading to technological determinism, the idea that innovation by its very nature generates specific social forms and values. The question for the Victorians and for us is this: does technological innovation determine the shape of society or do society and its values control technological innovation? Many early Victorians believed that mechanized textile production, because of its inherent nature, demanded long hours and dangerous conditions for textile workers. Many Americans believe that the technology of cars demands more road building. Such disregard of social and cultural systems makes us forget that society has the power of choice. Victorian mill owners finally did agree to the limitation of working hours and the fencing of dangerous machinery. And in our time, cities have regulated the number of cars entering urban areas. The story of Victorian technology, then, lies in both the narrative of invention and innovation in the development of machines and the account of transformations in the social and cultural value systems.

1

The Factory System

From the opening of the Liverpool and Manchester Railway in 1830, we can look forward and backward. A look backward sees the origins of steam power and the railway in the eighteenth century. A look forward sees the steam engine and the railway as crucial components of the complex structure of mass production that emerged in the early Victorian period. It is this factory system that transformed Victorian England and the world.

THE STEAM ENGINE, THE AGE OF STEAM

Let us begin with the steam engine, the dominant mechanical device that powered the water-pumping machinery that kept the coal mines dry, the railways that crossed England, the factories of Midland industrial cities, and the steamships transporting raw materials and manufactured goods across the oceans. The early-Victorian age can indeed be called the age of steam.

It would be convenient to say that the exemplary machine of the Victorian age was invented in the Victorian age, but that would be to simplify the process of technological innovation. Victorian technology did not mark a sudden break with the past but, especially in its early phases, a gradual process of improving mechanical devices developed in the preceding century. This is especially true of the steam engine and the steam locomotive.

The steam engine is essentially a mechanical device in which heat converts liquid water to vapor. The heat energy of this expanding steam under the pressure of containment is then released in a controlled fashion to be converted to mechanical energy that can move material objects, for example by pushing a piston in a cylinder. In searching for the origins of this machine, we can look to Hero of Alexandria, a Greek polymath who in the first century BCE boiled water in a bronze sphere he called an aeolipile so that the steam emerging through outlets caused the sphere to rotate. Such a device was for the Greeks only a curiosity, not a useful source of power.

Work toward the application of steam to industrial power in England begins in the first decades of the eighteenth century with the work of Thomas Newcomen, a blacksmith who experimented for ten years to develop a stationary steam engine before building his first working engine in 1712. The term stationary is generally used to describe an engine set in place, for example to pump water from mines as did Newcomen's first engine, or to power machinery in a factory, in distinction to the engines used to propel railway trains or steamships. Such an engine was needed for the pumping of water from coal mines, quite specifically to provide a reliable source of power to move the large wooden beam that drives the water pump.

The Newcomen steam engine is of a kind called the atmospheric engine since its operation depends on the pressure of the atmosphere. If we begin with the heavy beam tipped down, the cylinder below the steam piston is filled with steam, thus keeping the piston raised. Water is then sprayed into the cylinder, condensing the steam. A vacuum results and the pressure of the atmosphere pushes the piston down, thus pulling the main pump piston up. At the bottom of the steam piston stroke, a valve opens allowing air to enter, restoring the steam cylinder to normal atmospheric pressure, and the heavy beam falls by gravity. Steam is introduced into the cylinder again, and the process repeats itself continuously.

The Newcomen atmospheric engine effectively supplanted animal power with steam power in eighteenth-century coal mines and remained the state-of-the-art engine for these indispensable mines into the early nineteenth century. Still the machine was very large, not wholly reliable, and relatively inefficient in transforming steam energy into mechanical energy.

The next important innovation, and it is important to stress innovation here, is associated with James Watt. There is a legend that the young James Watt looked at a steaming tea kettle on the fireplace in a cottage in Scotland and seeing the water vapor emerge powerfully from the spout said, "Let there be a steam engine." Such legends, centering on the idea of a Eureka moment, from the sense of the Greek word Eureka, "I have found it," locate technological change in the individual as inspired genius. The reality of invention is more prosaic.

James Watt was a craftsman, an instrument maker in a workshop at the University of Glasgow. There in 1763 he repaired a model Newcomen engine. After much experimentation he showed that about 80 percent of the heat of the steam was consumed in heating the cylinder. Because the steam in the cylinder had to be condensed by an injected stream of cold water in order for the piston to fall, the cylinder itself was cooled in each repetition. Thus the cylinder had to be reheated with each cycle.

Watt's critical innovation to the atmospheric engine was to add a second separate cylinder called the condenser. To create a vacuum, steam from the heated cylinder flowed through a valve into this separate cylinder where cold water cooled and thereby condensed the steam. The great advantage of adding the condenser was that the temperature of the cylinder that contained the heated steam did not fall on each cycle. Heating this cylinder on each repetition with the resultant waste of energy was no longer necessary. It is the Watt steam engine with the separate condenser that became the dominant form of the steam engine powering locomotives and looms in the early nineteenth century.

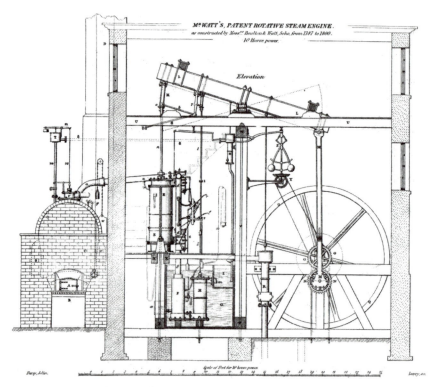

Watt steam engine. The 1787 drawing is captioned "Mr Watt's Patent Rotative Steam Engine as constructed by Messrs. Boulton & Watt, Soho. 10 Horse power." This technical drawing shows the boiler for creating steam, the steam cylinder containing a piston that moves the beam vertically, the water-cooled steam condenser, the steam-engine governor with its two rotating arms ending in heavy balls, and the wheel made to rotate by the geared vertical action of the beam. [Reprinted with permission of Science Museum/Science & Society Picture Library.]

In 1765, Watt had a small working model. Soon he had a functioning full-scale prototype built. Here we see the necessity for the innovator to work with others. The new Watt engine depended on the skill of craftsmen who could build the cylinders with tolerances that allowed the piston to move up and down and yet not allow steam to escape. It must be emphasized that innovation required not only experimenters like Newcomen and Watt, whose names have come down to us, but also the anonymous artisans who could do the demanding work of fabricating the sturdy-yet-delicate machines.

The manufacture of steam engines also needed financial capital, what we now call venture capital, to support the development of the machines until the company could turn a profit. In 1773, Watt, the university based instrument worker, formed a partnership with Matthew Boulton, who helped provide the capital to produce the Watt engine. The firm of Boulton & Watt soon became the leading firm of manufacturers and its name became synonymous with the steam engine. At first the engines were used to pump water from mines. Then, as a rotary gearing was developed, the stationary engines were used as the source of power in textile mills. The availability of reliable Boulton & Watt engines enabled the development of the railways in the 1830s and 1840s. Conversely, the expansion of the railway with its new demand for steam locomotive engines brought an enormous financial boom to the firm of Boulton & Watt.

THE STEAM-ENGINE GOVERNOR AND FEEDBACK MECHANISMS

The reliability of Watt engines was enhanced by Watt's invention of one of the most important devices of the industrial revolution, the steam-engine governor that brought the self-regulation of the enormous but variable heat energy of the engine. The governor was created to solve the specific problem of keeping the entry of steam to the engine consistent so that the engine could operate at a steady speed. For the engine to provide a steady input of power to a mill, the steam emerging from the boiler had to be of the proper strength, powerful enough to set in motion the upward moving piston that drove the rods to which the power looms were connected. If the expanding steam dropped below a minimum, the engine would not provide sufficient power. If the expanding steam became too powerful, the engine would run too quickly and self-destruct. It became clear that for steam to provide power efficiently and safely to the mills, some mechanical device was necessary to regulate the continuously varying pressure of the steam within the boiler.

As with much in the history of nineteenth-century invention, the device that answered this pressing need was developed in the late eighteenth century. James Watt's steam-engine governor, developed in the 1780s, provided a simple feedback mechanism by which variations in speed of the engine continuously feed back a signal that controls the input of steam to the engine in order to maintain a stable state. Watt's governor, the basic form of which continues in use to the present day, consists of two arms with a heavy ball attached to the end of each arm. These arms are attached on hinges to a single column that is rotated by the action of the steam engine. These balls are attached to a valve that controls the flow of steam to the engine. As the speed of the engine increases, the column turns more quickly and centrifugal force moves the hinged arms and the attached balls outward, thus narrowing the opening of the valve and reducing the amount of steam entering the engine. As the power of the engine drops, the speed of the engine is reduced, the central column rotates more slowly, the hinged arms and balls drop, the valve to the engine is opened more widely, and more steam enters the engine. As the speed of the engine then increases, the hinged arms again rotate and the balls move upward, reducing the entry of steam.

Thus the feedback loop continues in a permanent and fully automatic process of adjustment. The governor governs or rules the machine without the need of human supervision. As valuable as the Watt steam-engine governor was to the first stationary steam engines, the device was particularly valuable to the steam-powered locomotive as a moving steam engine whose speeds must increase and decrease as the train stops at a station, then attains traction to begin motion.

The governor is historically important as the exemplar of devices that can run consistently and efficiently by adjusting themselves to varying conditions without human intervention. Such automatic or in Victorian terms self-regulating devices were as crucial to the advance of Victorian technology as to the increasing industrial automation in our own world. Self-regulating machines operate on the general principle of feedback, the process in which the output of a system generates a signal that is fed back to control the system. As internal and external conditions change, the signal operates in a constant feedback loop to control the ongoing behavior of the system and thus maintain the stability of the system. In the nineteenth century, feedback mechanisms enabled the increasing power of the steam engine for the factory as well as the locomotive and steamboat by guaranteeing consistent, regular, and yet safe operation.

In addition to the governor, the very nature of the steam engine also demanded a feedback device to prevent the most common of Victorian accidents, steam boiler explosions. The steam engine operates essentially through heating water to generate the expansive energy of steam within the confined space of a boiler before the steam enters the engine. But building up a head of steam poses a serious danger, for if the steam becomes too expansive it can rupture the containing walls of the boiler. The early history of the stationary engine is replete with boiler explosions, as is the history of the steamship and the steam locomotive. Such accidents were frequent in the early days of the steam locomotive when water was heated in the boiler under high pressure in order to reach high speeds. Such railway boiler explosions were particularly destructive since they often occurred when the train was moving at high speed. The accident would stop the train suddenly, often derailing the cars behind, resulting in injury and death for the passengers.

Since many of these boiler explosions were caused by the incompetence or inattention of human attendants who allowed the pressure to exceed safety limits, an automatic device was needed to supersede the all-too-human limitations. This simple feedback device became known as the safety valve or relief valve. If pressure in the boiler grows too high, a valve opens to provide relief by allowing the excess steam to exit, and thus prevent explosion. In a continuous feedback loop, as steam pressure in the boiler eases, the valve closes, allowing pressure to build again. The earliest safety valve used a weight calibrated to a certain pressure to maintain the power of the steam. When the steam pressure reached a certain point, the weight would retract, releasing the steam. These weight-activated valves had a problem in that they could be activated when the engine hit a bump in the track. To replace these early devices, a valve was developed that employed a spring to deal with sudden accelerations. The installation of safety valves on the sides of the steam locomotive boiler did much to reduce devastating boiler accidents on railways. The functioning Victorian steam engine, then, must be seen not as a single device, but as a system incorporating the burning of fuel, the expansion of steam under pressure, and the conversion of pressure to mechanical motion, all self-regulated by automatic or feedback mechanisms.

The Victorians recognized that mechanical self-regulation, exemplified in the steam-engine governor and the steam-engine safety valve, was the enabling principle for progress toward large-scale, rapid, and continuous production in the factory. Charles Babbage, the inventor of the Victorian proto-computer, found a fusion of aesthetics and efficiency in "that beautiful contrivance, the governor of the steam-engine ... Whenever the increased speed of the engine would lead to injurious or dangerous consequences, this [the governor] is applied; and it is equally the regulator of the water-wheel which drives a spinning-jenny."[1] Andrew Ure, an influential Victorian defender of industrialism, praises the self-acting machinery of the ideal factory where "a self-acting governor attached to each wheel adjusts its velocity to the purposes of the factory, and is never in a state of repose, but is seen incessantly tightening or slacking the reins of the mill-gearing, so to speak, according to the number of machines moving within, and the force of the stream acting without."[2] Ure himself was proud of inventing what he called "the heat-governor, or thermostat, [that] would furnish the factory proprietors with a self-acting means of regulating the temperature of their apartments."[3]

The development of the steam engine displays several themes that run through the history of nineteenth-century innovation. First, Newcomen was a blacksmith, Watt a skilled craftsman at a university instrument workshop. Each, then, was an accomplished artisan who improved the steam engine through an extensive process of trial and error. In some contrast to our own times, new technology was not developed through what we call research by a university-trained elite of scientists or engineers in a specialized environment such as a university or corporate research institute. Rather, in the early decades of the technological revolution, invention arose from the craft tradition in such trades as blacksmithing and metal fabrication. Typically nameless craftsmen continued to be essential to the ongoing process of innovation through the nineteenth century.

Furthermore, innovation did not emerge from theory, but rather from the need to accomplish specific practical tasks. The advance of one technology necessitated advances in related technologies. The development of deep mining forced the development of the steam engine. As the need for coal grew, mines moved further beneath the surface and groundwater seeped into the tunnels. Animal power no longer provided sufficient force to work the pumps that drained water from the mines. Reliable mechanical power was needed. The Newcomen engine and Watt's improvements to that engine provided this reliable and efficient source of power. The technological progress continued as the Watt engine made possible the railway and the steam-powered mill.

THE STEAM ENGINE ON WHEELS: THE RAILWAY

The next crucial innovation in creating the age of steam was setting the steam engine on wheels. The Rocket, pulling carriages at the opening of the Liverpool and Manchester Railway in 1830, marked the advent of a new era.

As with the harnessing of the expansive energy of water vapor in the efficient Watt stationary engine, the use of steam to power forward motion was the result of innovations going back to the eighteenth century and brought together in the steam locomotive for the railway. In eighteenth-century mines, cars loaded with coal had been pulled along tunnels by a cable moved by a stationary steam engine. The crucial step toward the railway locomotive came with the innovation of moving the stationary engine onto the wheeled car itself to create a self-propelled vehicle in which the expansive power of steam was directed to turning the wheels. Such a vehicle could then haul cars behind it. The steam locomotives that preceded the Rocket manifest this crucial advance to self-propulsion. In 1804, Richard Trevithick built the first steam railroad locomotive. It hauled cars on iron tracks to carry workers at an ironworks. In 1808, he built a circular track in London that offered trips in carriages pulled by a steam locomotive appropriately called Catch Me Who Can. The first railway that carried passengers and freight was the Stockton and Darlington Railway, opened in 1825 on a rail line of seventeen miles. However, horses as well as steam locomotives were employed.

At its opening in 1830, the Liverpool and Manchester Railway brought these earlier innovations into the system we call the railway. The horse was wholly replaced by the steam locomotive, although the energy of the new machine was measured in units called horsepower. George Stephenson, builder of the Rocket, solved the crucial engineering problem of converting the heat energy of steam to a functional form of mechanical energy. Specifically, he designed gears to transfer the up-and-down motions of the piston driven by the expanding steam coming from the boiler into rotary power to turn the iron wheels. He did this by coupling the piston rods moved by the steam to connecting rods that drove two large front wheels. The steam was generated in the boiler section of the locomotive. The boiler was fed coal through a hatch and the fire was tended by the stoker. The steam locomotive had to carry its own coal supply in a separate car, the tender. This engine, as locomotives came to be called, eventually ran at the then astonishing speed of thirty miles per hour.

The Liverpool and Manchester Railway brought into being the technological system of rail transport followed to this day. All power was supplied by steam. The line carried both passengers and freight. The trains ran on iron tracks or rails. There were two tracks, one eastbound and one westbound. Stations were built along the line for goods and for passengers. The trains ran according to a timetable. Fares were charged based on the distance traveled. The fares were graded into first, second, and third class, thus opening rail travel to all at reasonable prices and providing a range of freedom for the ordinary person.

After the opening of the Liverpool and Manchester railway, the expansion of the railway system was explosive. In 1830, there were approximately one hundred miles of railway track. The 1840s saw the great railway boom as investors in a burst of free enterprise formed joint stock companies, some honest and some corrupt, that proposed a multitude of railway companies. Some of these plans succeeded, some did not. Yet by 1852 there were 6,600 miles of track in England. At the end of Victoria's reign, the total mileage of the system had reached almost nineteen thousand. The Victorian railway system was arranged as trunk or main lines with secondary lines reaching smaller towns. Primarily, the trunk lines connected London to the industrial cities of Birmingham and Manchester, to the ports on the English Channel, and to Scotland. A line was also built from London to the port of Bristol on the west coast of England where a ferry linked England to Ireland. From the beginning, railways connected the industrial cities such as Manchester to the seaports such as Liverpool, from where ships carried manufactured textiles to the Americas and the Empire.

As railway lines came to cover almost all of England the image of the train marked by spouting steam moving over the immemorial pastoral landscape as the "machine in the garden"[4] became the dominant image of the transformation and the conflict of the time. The moving engine speeding across fields and valleys on newly built viaducts and crossing rivers on iron bridges seemed at war with the beauty of an ancient land. In 1844, William Wordsworth, appalled at the incursion of the railway into the Lake District, the mountainous area that inspired his romantic nature poetry, in his sonnet, "On the Projected Kendal and Windermere Railway," asked:

Is then no nook of English ground secure
From rash assault? Schemes of retirement sown
In youth, and 'mid the busy world kept pure
As when their earliest flowers of hope were blown,
Must perish?[5]

This new machine, the steam locomotive, was only one element of a technological system. The dramatic spread of the railway across pastoral England in the early decades of the Victorian age necessitated a vast construction project akin to the building of the Interstate Highway System in America in the latter twentieth century. The endeavor involved new technologies as well as the traditional power of animals and human muscle. For the locomotive to move efficiently along its iron rails the roadbed had to be as level as possible, with an incline for hills of no more than one foot in thirty-six linear feet. In low-lying areas the ground had to be built up into embankments of stone and earth to carry the tracks. Furthermore, since it was most effective for the roadbed to move in a straight line, through a hill rather than around it, building the railway necessitated enormous excavations that reshaped the English landscape during the 1840s and beyond. There were cuttings, excavations through the sides of hills so the roadbed would run level. One cutting on the Newcastle and Carlisle line was a mile long, reaching a depth of 110 feet. And where the ground was too steep for cuttings, long tunnels were drilled through the hills themselves. The experience for such tunneling and digging had been gained through the long British experience of mining, but the scale of this early-Victorian earth-moving was unprecedented. Thousands of laborers, called navvies, moved the earth using pick and shovel, horse-drawn carts, explosives, and eventually, steam-powered digging machinery.

The valleys were crossed by great viaducts of stone, brick, timber, and iron so the track could run level across empty space. These great spans remain as attractive to contemporary viewers as they did to the Victorians. Even Wordsworth felt that the literary imagination could draw inspiration from these structures. In his sonnet of 1833, "Steamboats, Viaducts, and Railways," he asserts that "Nature doth embrace/ Her lawful offspring in Man's art."[6]

Of equal importance to the excavation, and to some the destruction, of the English countryside was the disruption caused by the coming of the railway to the cities, especially London. As the railway system developed as a web of tracks connecting London with the furthest reaches of the country, each railway company had to construct a terminus or large rail station in the capital. The construction of viaducts, cuttings, and tunnels in the capital caused enormous devastation. Charles Dickens's account in his novel, *Dombey and Son*, evokes the confusion and also the energy in the sudden eradication of old worlds, what we call in our own time the destruction of urban neighborhoods, by the coming of the railway to London. Here Dickens describes the demolishing begun in 1834 of the Camden Town section of London for the right-of-way to Euston Station, the terminus of the London-Birmingham railway line. Speaking for the age, Dickens registers not only the physical destruction of housing, but also the breaking up of the pre-industrial social cohesion of established custom:

> The first shock of a great earthquake had ... rent the whole neighbour-hood to its centre. Traces of its course were visible on every side. Houses were knocked down; streets broken through and stopped; deep pits and trenches dug in the ground; enormous heaps of earth and clay thrown up; buildings that were undermined and shaking, propped by great beams of wood. Here, a chaos of carts, overthrown and jumbled together, lay topsy-turvy at the bottom of a steep unnatural hill; there, confused treasures of iron soaked and rusted in something that had accidentally become a pond.... Boiling water hissed and heaved within dilapidated walls; whence, also, the glare and roar of flames came issuing forth; and mounds of ashes blocked up rights of way, and wholly changed the law and custom of neighbourhood.[7]

If the railway disturbed many Victorians through its intrusion into the English landscape and its tearing the fabric of urban life, other Victorians saw in the locomotive and train itself an inspiring emblem for the transformative energy of the age. The vision of a disembodied energy seemingly embodied in the machine is eternally captured in Joseph Mallord William Turner's painting of 1844, *Rain, Steam and Speed—the Great Western Railway*.[8] In this representation of a speeding locomotive driving over a viaduct through a rainstorm, cities have disappeared as does the land itself. It is as if the locomotive has leapt the bounds of earth to become pure spirit. As the title suggests, the natural element of rain is equated with the man-made energy of steam in the locomotive. Human inventors have not only liberated the power of nature in the steam locomotive, but also fused the iron machine with the elemental forces of nature, with wind and rain.

STANDARDIZATION AND THE RAILWAY SYSTEM: TRACKS AND TIME

The use of rails to speed self-propelled vehicles seems inevitable to us, but, as with much in the history of technology, what is apparent in the present did not seem self-evident in the past. In eighteenth-century mines, wagons filled with coal were drawn to the surface by cables attached to a stationary engine that ran along metal plates rather than rails. The eighteenth century also saw steam-powered land coaches that were designed to run along the network of coaching roads used by horse-drawn stage coaches. In the early nineteenth century, Richard Trevithick experimented with steam road coaches. Unfortunately, these steam vehicles had a propensity to overturn due to the poor road surfaces of the time. The use of iron tracks for steam-powered vehicles was assured with the opening in 1825 of the Stockton and Darlington Railway on which a steam locomotive drew cars along a set of iron rails.

The rapid development of the railway system after the great success of the Liverpool and Manchester Railway led to the first great battle over technological standardization, the gauge controversy. To create the railway tracks, iron rails were laid on ties, or in the British term, sleepers. For the efficiency of mass production, track builders and locomotive manufacturers had to work according to a consistent uniform width or gauge between the rails. But what was this gauge to be? This question brought bitter rivalries among great commercial interests.

To understand the intensity and the stakes in the issue of railway gauge, consider the significance of universally accepted standards in technological innovation and its application. In our day we can see the need for standard specifications in electronic equipment, media players, and phones. And we witness enormous conflicts over what the standards are to be as well as the significant economic consequences for corporations whose phone or media player does not fit the standards that win out. So it was with the fight over a standard gauge for the swiftly developing Victorian railway system.

In the 1840s, separate companies were building separate parts of the railway network and each company used a different gauge. The Liverpool and Manchester Railway used the so-called narrow gauge of 4 feet 8½ inches, based in vestigial fashion on the average width between the wheels of a horse-drawn coach. However, the Great Western Railway employed the broad gauge of 7 feet for their tracks. Other companies used different widths. Since locomotives were necessarily built to run on specific gauges, trains could not run on the tracks of other companies. Passengers had to change trains at break-of-gauge points.

Only by setting a standard, uniform gauge could England create a truly national railway system in which the trains of one line could run on the tracks of another. Serious economic interests were involved since the company whose gauge was accepted would become dominant. The Great Western, for example, would control the system if the broad gauge was adopted. Parliament investigated the issue through a Royal Commission, but in the general Victorian spirit that government should not interfere in commerce, refused to set a standard. It was only through competition between companies and the general understanding that a standard system was beneficial to all that narrow gauge finally won out.

Adopting a standard gauge was crucial for several reasons. For industrial manufacturing, as always, the principle of standardization reigned. Locomotive manufacturers could attain efficiency in their mass production of engines only if their product was uniform. Rather than producing locomotives with differing wheel bases for different gauges, the factories could design their machine tools for a standard engine.

The gauge wars of the 1840s illustrate an inherent logic in mechanization. The expansion of mechanical systems from workshop to region, from region to nation, and then to the world necessitates that the local idea must give way to an imposed consistency created by government or by the needs of the system itself. More generally, as the nineteenth century demonstrates, innovation may proceed from individual actions, but an industrial system can only be consolidated if standards are established to provide a context in which further innovation can take place.

The intensely rapid pace of the railway in the 1830s and 1840s created, as we have seen, a number of disruptions, not the least of which was the inability of the literary imagination to keep pace with the swift technological change, a problem seen also in our own time. In 1837–1838, the young Alfred Tennyson, later to become Poet Laureate, wrote "Locksley Hall," a poem in which the narrator overcomes his grief at an unrequited love by identifying with the material progress of England:

> Not in vain the distance beacons. Forward, forward let us range,
> Let the great world spin forever down the ringing grooves of change.[9]

Although the young Tennyson had sufficiently absorbed the idea of the railway into his imagination so that he could employ the railway tracks', "ringing grooves," as a figure for technological progress, he did get his facts wrong. In one of the number of standardizing decisions for the railway system, the top surface of the rails was rounded. The wheels that ran on these rails were flanged with a projection on the outer edge of the wheel to keep the train on the track. On his first train trip Tennyson had simply not looked at the mechanism and assumed that the wheels ran on grooved tracks.

The railway system flourished by applying the industrial principle of standardization not only to its functioning material elements, but to time itself. As the width between the rails and the rolling stock itself became standardized with the expansion of the railway system, so did the non-material element through which the railway ran, time itself. It may be difficult for us to imagine that in England before the railway each locality kept to a different time. Local time as called out by the church bells was essentially sun time, calculated from noon when the sun was directly overhead. And no one cared; the difference in time made no difference for ordinary life in an isolated, self-contained community. The days of the week were universal, but it mattered little, if at all, if it was three o'clock in Norfolk and 4:15 in London. After all, communication by horse between towns and cities was measured in days, not in minutes. And the horse-drawn stage-coaches ran on a schedule that was approximate at best.

As in the case of the track gauge, it soon became clear that a national standard time had to be established to enable the smooth and safe running of a national railway system. For the train rather than the stagecoach, the timing of transportation had to be predicated with accuracy reliable to the minute. There was the matter of safety. Trains had to be scheduled to move along the tracks at regular intervals so as not to collide. The railway gave the gift of mobility, but to use this gift people had to know the exact arrival and departure times of trains. Thus the timetable was born. The new national standard time was measured not by the sun, but by the crucial machine of industrialization, the clock.[10] In 1850, the railways agreed to observe one time nationwide, rather than the local time at the individual stations. By convention, the time was set at Greenwich Royal Observatory in Greenwich, England. The use of such coordinated time was made possible by yet another invention, the electric telegraph that could virtually instantaneously provide a time check throughout England and Ireland. This standardized time under which we still live, Greenwich Mean Time, was originally called railway time.

In his novel of 1907, *The Secret Agent*, Joseph Conrad imagines a group of anarchists, terrorists in contemporary parlance, who seek to bring England to chaos and confusion by blowing up the Greenwich Observatory and thus destroying Greenwich Mean Time. The explosion does not take place, and the goal of destroying the industrial world by destroying its time is intended by the author to show the folly of the anarchist mind. Yet the fantasy of the anarchists does draw upon our sense that modern technological systems do depend on an invisible set of agreed-upon parameters, such as the industrialization of time itself.

AUTOMATIC STEAM-POWERED TEXTILE MACHINERY

The making of finished textiles from plant and animal fiber is among the most ancient of human activities. The process can be divided into two elements. The first is spinning, the transformation of raw material such as wool or cotton fiber into threads. We are all familiar with the spinning wheel, in which the large wheel powered by the foot pedal twists raw wool into yarn gathered on the small spindle. Indeed, this foot-powered device has come to emblemize traditional, pre-industrial life. The wheel is typically turned by a woman. The setting is the home. The paragon of female domesticity spins yarn only for her family's needs rather than for sale to others.

The other element of textile-making is weaving, the transforming of threads into fabric. For weaving, the basic element, unchanged over centuries, is the loom, like the spinning wheel a traditional technology. Basically, the loom consists of a stable frame to which is attached the warp, a fixed set of threads. Through the warp the shuttle, originally moved by hand, pulls crosswise another set of threads, the woof threads, interlacing over and under the warp to create a finished fabric.

In pre-industrial England, weaving was primarily done by men known as hand-loom weavers. As the name implies, the work at the loom was done without the use of an external power source. The weavers drew the weft through the warp by hand and the movement of the frame to push the thread tightly together was powered by a foot treadle. The work was done by men rather than women. It was domestic in the sense that weavers usually worked in their own homes, but not domestic as in the sense that the weavings were sold beyond the household. In this system jobbers would usually bring yarn and thread to the individual weavers who would weave the thread into cloth. The jobbers would then return and pay the weavers for the finished fabric. The work was proto-capitalist in that weavers did not keep the products of their labor, but rather produced commodities to be sold finally for the profit of others. It was the hand-loom weavers who suffered as much as any other group from the revolutionary industrialization of textile production.

The transformation of these ancient tasks of spinning and weaving by the application of steam power to automatic machinery and the movement of production from the home workshop to the factory marks the first major shift to mechanized manufacturing in the nineteenth century.

As with the steam engine and the steam locomotive, the development of efficient automatic steam-powered textile machinery can be traced to the late eighteenth century. We can begin with the invention by John Kay in 1733 of the flying shuttle, the first major device in the automation of textile production. In hand-loom weaving, the width of the fabric was limited by the reach of the weaver who had to pass the shuttle carrying the weft threads through the warp. Furthermore, the slowness of the motion of the hand limited production. In the flying shuttle the weaver can jerk a cord sending the shuttle flying with the woof threads across the warp and with another motion of the cord bring the shuttle flying back. This seemingly simple device had the effect of increasing the width of cotton cloth and the speed of production of a single weaver at a loom.

The higher rate of textile production with the flying shuttle generated an increased demand for spun yarn and thus encouraged the invention of newer machines to fulfill this need. Here appears Richard Arkwright, who was to enter the Victorian pantheon of the inventor as hero. From earlier spinning machines, Arkwright constructed an improved machine called the water frame, which he patented in 1769. In this machine, driven by a water wheel external to the workshop, a bundle of untwisted but parallel wool fibers called a roving was pulled through a series of rollers rotating at varied speeds. This process drew the fiber into a thin thread. The thread was passed on to a flyer or rotating frame that twisted the wool fiber to give it strength, then gathered the twisted wool thread on a wooden cylinder or bobbin. To facilitate the spinning of the water frame, Arkwright developed a carding engine patented in 1775 consisting of a set of drums with metal teeth that drew the raw bundle of cotton or wool into parallel fibers.

In the 1760s, another mechanized invention for spinning appeared, the spinning jenny, whose invention is generally credited to James Hargreaves. This water-driven machine also multiplied the amount of thread that one person could spin. Basically, the spinning jenny increased the action of the traditional spinning wheel by increasing the size of the wheel and setting the wheel on its side. The wheel fed eight spindles. A set of eight rovings were attached to a beam that could roll from the spindle end to the wheel end on a horizontal frame. The operator could roll the beam back and forth over the yarn to draw it out to the proper thickness. A clamp-like device in the roving beam allowed the operator to then release all the threads at once, to be collected on spools.

The acceleration in the efficiency of spinning with Hargreaves's and Arkwright's inventions created a demand for new machines for weaving. In 1785, Edmund Cartwright patented the power loom, precisely in response to this need. In his own words:

> Happening to be at Matlock, in the summer of 1784, I fell in company with some gentlemen of Manchester, when the conversation turned on Arkwright's spinning machinery. One of the company observed, that as soon as Arkwright's patent expired, so many mills would be erected, and so much cotton spun, that hands never could be found to weave it. To this observation I replied that Arkwright must then set his wits to work to invent a weaving mill.[11]

But first, Cartwright developed a new form of the loom. His early design was rather crude. This device followed the traditional operations of the loom, but in this machine the shuttle operated by mechanical rather than by human power. The loom was originally run by water power, but could be modified to use the new steam engines. The original design was gradually improved. One problem was that adjustments to the warp necessitated the closing down of the loom. In 1803, Thomas Johnson invented the dressing frame, which enabled power looms to operate continuously, thus multiplying their efficiency. As a Victorian historian of the industry noted with pride, "Before the invention of the Dressing Frame, one Weaver was required to each Steam Loom. At present a boy or girl, fourteen or fifteen years of age, can manage two Steam Looms, and with their help can weave three and a half times as much cloth as the best hand Weaver."[12]

And Arkwright did "invent a weaving mill" as Cartwright predicted. In 1772, Arkwright opened what can be considered the first factory, a large brick textile mill five stories high at Cromford in the North of England. This was a spacious structure where the raw material could be delivered and many people could work simultaneously at spinning and weaving. In this mill, the entire process of textile production was centralized. Importantly, the making of textiles was divided into separate stages that could each be undertaken by specialized large-scale machines.

As the practice of weaving with power looms became automatic, the setting or, in our terms, programming of the loom became more sophisticated. In 1801, Joseph M. Jacquard of France invented what came to be called the jacquard loom, a device quickly adopted by British mill owners. The jacquard had the advantage of being able to weave automatically and repetitively into fabrics the interlacing designs of flowers and animals so beloved by Victorians.

In the jacquard loom, the fabric designer translates his or her textile patterns to be woven for the power loom by punching holes into paper cards. In contemporary terms, the design is coded into a program that is translated into holes in a series of punch cards, similar to the punch cards of early IBM computer systems. The holes physically allow wires in the loom to raise and lower different warp threads in a complex sequence within a fully automated weaving operation: "the pattern of holes ... dictated which coloured threads were lifted and which were not, so that the desired pattern was woven."[13] This punch card device for the automatic running of a power loom was to provide the Victorian inventor Charles Babbage with the inspiration for his proto-computers.

The mass production of textiles also called for automatic printing of colored designs on the finished cloth. Here the Victorians invented complex automatic machines that would repetitively print the same pattern in many colors without human intervention. The self-acting fabric printing machine

Spinning machines in textile mill. A man and a woman operate two efficient spinning machines or mules that move back and forth on tracks. There are multiple bobbins for the finished yarn. Power is transmitted to the mules by belts connected to the rotating rods overhead powered by a stationary steam engine outside the mill. From Edward Baines, *History of the Cotton Manufacture in Great Britain*. London: H. Fisher & Co., 1835. [Reprinted with permission of the British Library Board - Shelfmark 1044.g.23, opposite 211.]

was praised by Andrew Ure, one of the chief Victorian champions of mechanization: "The finest model of an automatic manufacture of *mixed* chemistry is the five-coloured calico machine, which continuous and spontaneously, so to speak, prints beautiful webs of cloth with admirable precision and speed."[14] In the pre-industrial world, vivid dyes for fabrics had been produced from natural materials, necessarily resulting in an attractive unevenness in color. The success of mechanized printing depended on standardization of color. With the application of the science of chemistry, Victorian technology was able to produce artificial colors that if more regular were less attractive.

CAST IRON

The boilers of the steam engine, the rails of the railway, the grand railway bridges, the columns and joists of the textile mills—these elements of the Victorian industrial system were fashioned by yet another technology, the manufacture of strong, reliable, standardized cast-iron components. Again, this innovation had its roots in the later eighteenth century and was applied to large-scale mass production in the nineteenth.

Like spinning and weaving, the making and forging of metals has been central to human progress, as seen in the names that characterize historical periods—the Stone Age, the Bronze Age, the Iron Age. Indeed, the Victorian period might be termed the Cast Iron Age. By the eighteenth century, the use of such metals as bronze, iron, and steel had a high degree of refinement, but such metals were produced in small workshops, forges, and foundries. Metal was therefore expensive and used where its hardness was necessary as in weapons, tools, and nails. Nostalgia for pre-industrial times looks to the blacksmith shop, where shoes were fitted to horses and small tools manufactured and repaired. The blacksmith Joe Gargery of Charles Dickens's *Great Expectations* was the emblem to the Victorians of the heroic craftsman whose activity was already outmoded. In Joe's forge attached to the home, as in other metal workshops, a lump of heated metal had to be hammered into shape by hand tools producing what is called wrought iron.

For centuries, iron used in these workshops was produced through the process of smelting or heating, often using a blast furnace. In the pre-industrial blast furnace, the combustion material and ore were supplied from the top while an air flow was supplied from the bottom of the chamber. The fuel most often used for this process which demands high and consistent temperatures had been charcoal.

The basic change in this process, one of the essential innovations of the industrial revolution, took place in the eighteenth century. In 1709, at his foundry in the small town of Coalbrookdale in the Severn River Valley in the Midlands, Abraham Darby developed a blast furnace that replaced the use of charcoal for heating with coke, a fuel made by refining coal. The use of coke allowed higher heat within the furnace. The coke-produced iron became far cheaper to produce than charcoal-made iron and Darby was soon supplying the nearby forges with bar iron produced from his blast furnaces. Expanded use of the innovative coke-fired blast furnace was enabled by the abundance of coal in the British Midlands.

Technological progress moved toward the mass production of cast iron. One important development was the change to hot blast, patented by James Beaumont Neilson in Scotland in 1828. In an innovation reminiscent of the addition of a condenser to the Watt steam engine, a stove as large as the furnace next to it was added into which the waste gas from the furnace was directed and burnt. The resultant heat was used to preheat the air blown into the furnace. The elimination of reheating reduced production costs.

In the improved blast furnaces of mid-Victorian times, coke, limestone, and iron ore were poured in the top. Air was blown in through tubes near to the base. The blast of air allowed the combustion of the fuel at the base. This reduced the iron ore to the pure metal, which being heavier sank to the bottom of the furnace. The unwanted silicon and other impurities are lighter than the molten iron, and thus rose to the top. The iron then flowed out of the bottom of the blast furnace into molds. This metal is called pig iron. Further processing was needed to reduce the carbon content to create a suitably hard cast iron for tools or for iron columns in buildings. The most successful new process used through the mid-century was puddling, in which iron rust was added to the molten pig iron and stirred by skilled workmen called puddlers. In this process the carbon impurities are oxidized to create a strong form of cast iron suitable for construction purposes.

The coke-fired blast furnaces of eighteenth-century Coalbrookdale were isolated entities within an agricultural setting and produced only small quantities of iron mostly used to cast pots or cauldrons. Only with the rise of industrial cities in the early nineteenth century did the blast furnace with its visible fires, bellowing smoke, and tall chimney become a common part of the urban landscape viewable by anyone who took the train from London to the Midlands.

IRON CONSTRUCTION

Strong under compression, inexpensive, and reliable in its composition, cast iron was ideal for construction purposes. Cast iron was the material with which the structures of an industrializing England were built. The use of iron for structural purposes began in the late 1770s when local businessmen commissioned Abraham Darby III, descendent of the originator of the Coalbrookdale coke-fired forges and himself a manager of foundries, to build a bridge across the Severn River using the plentiful cast iron from the Coalbrookdale blast furnaces. The Iron Bridge, as it was called, was completed in 1779. It is the first bridge in the world to use cast iron structurally. It was made part of a UNESCO World Heritage site to celebrate its centrality to the global industrial revolution.

As typical in the history of innovation, the Iron Bridge was called into being by need, in this case to provide easy passage across the Severn River because the area had become the busy scene of iron manufacturing. Although the architecture of the bridge with its arches and methods of joining the iron members relies on the traditional forms used in wood carpentry and stone masonry, the bridge takes advantage of the compression strength of cast iron and the ability to cast single long members of iron. The single arch spans one hundred feet over the Severn. Indicating the importance of prefabrication for iron architecture, each of the five arch ribs was cast in two halves and joined at the site. Although the Iron Bridge was originally constructed for foot traffic and horse-drawn vehicles, in its use of standardized components and of cast iron to support a long, uninterrupted span the bridge points ahead to the great construction projects of the railway age. The Iron Bridge is still standing.

Iron Bridge at Coalbrookdale. Completed in 1779, it is the first bridge in the world to use cast iron structurally. The bridge takes advantage of the compression strength of cast iron and the ability to cast single long members of iron. The single arch spans 100 feet over the River Severn. The cast-iron parts were prefabricated. The bridge is still in use. [Reprinted with permission of Science Museum/Science & Society Picture Library.]

The spread of the railway system necessitated the spanning of the waterways of Britain. The great railway bridges built from the late 1840s onward, descendents of the Iron Bridge, depended as much on the availability of cast-iron structural elements as on the genius of Victorian engineers.

With the spread of the railway system through the British Isles, there was increased traffic to Holyhead in Wales, the embarkation point for the ferry to Ireland. Creating a direct line from London to Holyhead demanded a bridge over the Menai Strait that divides Western Wales. The task was given to Robert Stephenson, one of the preeminent Victorian engineers and son of the Robert Stephenson who designed the Rocket for the Liverpool and Manchester Railway. Stephenson developed an innovative plan based on the structural strength of iron. He designed a bridge consisting of huge rectangular iron tubes made from riveted iron plates, one 230 and one 460 feet long and weighing about eighteen hundred tons each. Unlike earlier bridges, the Britannia Tubular Bridge, known as the Britannia Bridge, did not rest on an arch. Rather, the iron tubes ran as spans of an unprecedented length that rested on four masonry towers with some support from chains suspended from these towers. The stiffness, rigidity, and strength of the iron tubular construction enabled the bridge to bear the weight of loaded trains that ran inside these giant tubes. One of the many engineering problems to be solved was lifting the gigantic tubes into place. Stephenson resolved the question by floating the tubes to a point under their position in the bridge, then slowly raising the tubes by hydraulic lifts into place while setting masonry under them as they were lifted in case of a malfunction of the lifts. As a sign of the popularity of Victorian engineering as public spectacle, men and women watched from the shore as the floating tubes of the bridge were lifted onto their masonry supports. The bridge was completed in 1850. It is a testament to the brilliance of Victorian engineering that the Britannia Bridge continued in use until 1970 when an accidental fire weakened the iron so that the bridge had to be reconstructed.

With its use of iron construction, the Britannia Bridge became a model for other great Victorian railway bridges. Isambard Kingdom Brunel, another of the heroic Victorian engineers, used a similar technique in building the Royal Albert Bridge at Saltash in Cornwall, completed in 1859. Here Brunel suspended the iron roadway from rounded iron tubes above, an application of Stephenson's iron tubes. The Royal Albert Bridge manifests gracefulness in the curving iron tubes and the echoing curve of the suspension cables descending from these tubes to hold the roadway.

The potential of iron to span large spaces, as demonstrated in the arches of the Iron Bridge and the iron tubes of the Britannia Bridge, was put to use by the Victorians in another necessary structure of the railway system, the great train sheds of the major London railway stations. The example of the train sheds demonstrates how the availability of mass-produced cast iron made possible the industrial transformation of architecture. These still-standing monuments of Victorian engineering were, like most new types of structure in the machine age, built to fulfill a need generated by technological change. The problem was how to accommodate within the railway terminus in a single unobstructed space the huge number of tracks for trains beginning and ending their journeys. Furthermore, the space must provide platform access to thousands of passengers, protect passengers from the weather, and allow entry of natural light in the age before electricity. For the Victorian builders the solution was to create a new type of structure in a building utilizing the properties of cast iron and the methods of standardized mass production.

The train shed was essentially a structure supported by prefabricated iron. The space was spanned by a vaulting consisting of ribs of mass-produced, uniform iron parts. These ribs rested on similarly uniform iron columns. Between these ribs were set glass panes, now also mass-produced and standardized. The result was a vast, naturally lighted enclosed space that encompassed the busy life of trains and passengers. The terminus of the Great Western Railway at Paddington Station, completed in 1854 by Isambard Kingdom Brunel, had wrought iron arches in three spans, respectively

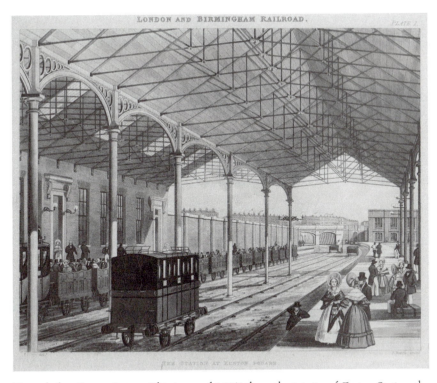

Train shed at Euston Station. This image of 1837 shows the interior of Euston Station, the London terminus of the London & Birmingham Railway. Passengers with luggage wait for their train; a steam locomotive carries passengers in the mix of closed and open carriages typical of early railway travel. Cast-iron columns, decorated arches, and iron struts support the roof with its glass skylights. Aquatint "The Station at Euston Square, London" by Thomas Talbot Bury, 1837. [Reprinted with permission of Science Museum/Science & Society Picture Library.]

spanning 68 feet, 102 feet, and 70 feet. The roof was 699 feet long. The iron and glass brought together by industrial methods was, however, decorated in forms that evoke the pre-industrial age. For example, where the iron roof ribs joined the iron columns, capitals in the Greek Doric or Ionic mode were often added.

LIGHTING BY COAL GAS

At the end of the eighteenth century, interiors were illuminated by candles or by lamps that burned whale oil. By the beginning of the nineteenth century, home and factory interiors were illuminated by the coal gas that was the byproduct of the coal mining that grew with industrialization.

As the mining of coal expanded at the end of the eighteenth century, miners increasingly encountered large quantities of the asphyxiating and explosive gas given off by the coal as it was excavated. But in the innovative spirit of the time, one person saw the possibility that this flammable property could be turned to productive ends. William Murdoch was an avid experimenter who worked for Matthew Boulton and James Watt at their Soho Foundry in Birmingham. He found that coal gas given off in the workshop from the distillation of coal was an effective source of light when set on fire under controlled conditions. In 1798, he used coal gas to light the main building of the Soho Foundry. By the early Victorian industrial decades, the advantages of coal-gas lighting were generally recognized. Gas lighting was 75 percent less expensive than candles or oil lamps. Soon, the large interior spaces of mills were brightly illuminated by gas lamps. The effect of gas lighting on the factory system was great. Hours of work could be extended since labor was no longer dependent on natural light or expensive candle power. Men, women, and children could now work effectively round the clock in the mills. Soon, gas lampposts, like those now preserved in our historic areas, were used for the illumination of public streets. The first use of gas street lighting was in London and came in 1807. Many English cities had followed by the 1850s. This illumination made the streets of Victorian cities safer, if not entirely safe. Gas was also piped to homes to feed small gas lamps. The improved illumination in the home was in part responsible for the great interest in reading during that period.

In the pattern of the times, manufacture, distribution, and profit were quickly centralized and industrialized. Gas works, essentially large factories for the production of gas, were established and became one of the visible markers of the industrial city. Here large amounts of coal were brought in and gas was produced by heating the coal in large furnaces. The gas was then, as now, distributed to businesses and homes through underground pipes. In 1812, the Westminster Gas Light and Coke Company was granted approval as the first company to supply London with coal gas. Eventually, the company laid 290 miles of pipes to supply its customers.

THE FACTORY AND THE FACTORY SYSTEM

The rise of the steam engine, automatic textile machinery, coke-powered blast furnaces producing iron, and coal-gas lighting all date to the late eighteenth century. In the Victorian age, it was the railway that brought these emergent technologies into a national industrial system. The rural workshops and foundries of the late eighteenth century and early decades of the nineteenth were isolated and separate. Their raw materials, fuel, and their finished goods moved slowly along the most efficient transport system of the time, an extensive network of canals. The enormous growth spurt of the railway network after the opening of the Liverpool and Manchester Railway demonstrates the pent-up demand for a unifying transportation mode.

Thus, the steam-driven railway was the key player in the emergence of mechanized industry. Trains drawn by steam locomotives ran on rails made of the iron from coke-fired furnaces. The railway brought iron ore to the blast furnaces and coal to the stationary steam engines that powered the factory. Rapidly, cheaply, and efficiently, trains carried American cotton from England's ports such as Liverpool to the textile mills of Midlands cities such as Manchester. From the mills, trains distributed the patterned cloth throughout England and to ports like Liverpool where steamships waited to take the finished textiles to America and through the British Empire, especially to the vast markets of India. Thus, with the coming of the railway in the 1830s, the first decade of Victoria's reign, innovations in machinery energized each other in an organization that from then to now has been known as the factory system.

The contemporary perception of steam-powered mechanized production within a factory as a dynamic structure is best expressed by Andrew Ure, the foremost British advocate of industrial capitalism, writing in 1835:

> The term *factory system*, in technology, designates the combined operation of many orders of work-people, adult and young, in tending with assiduous skill a series of productive machines continuously impelled by a central power. This definition includes such organizations as cotton-mills, flax-mills, silk-mills, woolen-mills and certain engineering works ... I conceive this title, in its strictest sense, involves the ideas of a vast automaton, composed of various mechanical and intellectual organs, acting in uninterrupted concert for the production of a common object, all of them subordinated to a self-regulating moving force.[15]

The Victorians realized that it was this new organization of workers and machinery that had created the sudden wealth of their society. The nineteenth century envisioned the binding of living workers and automatic machinery in the factory as a new form or social order, as itself an unprecedented form of the machine. In 1835, Edward Baines, Jr., another advocate of industrialization, looked back to the sources of industrialization in the eighteenth century. In his *History of the Cotton Manufacture of Great Britain*, he wrote of Arkwright not only as a man of "high inventive talent" in developing "spinning by rollers," but also as a person of

> unrivalled sagacity in estimating at their true value the mechanical contrivances of others, combining them [textile machines] together, perfecting them, arranging a complete series of machinery, and

constructing the factory system—itself a vast and admirable machine, which has been the source of great wealth, both to individuals and to the nation.[16]

Karl Marx also saw the factory as an immense cyborgian entity of machines fused with human beings imagining "what we may call the body of the factory, i.e., machinery organised into a system."[17] To the technological imagination of James Nasmyth, famous for his invention of the steam hammer, in the great industrial city of Manchester, the "whole of the buildings, howsoever extensive and apparently complicated, worked like one grand and perfectly constructed machine"[18]

The factory considered-as-machine comprises an unprecedented system. It is at once a new type of building; a site powered by steam: a set of interdependent automatic machines; a social order of workers as organized machine tenders; and a workshop connected domestically by the web of railways and globally by steamships to sources of raw materials and to buyers of its products. The next sections will consider these components separately, while keeping in mind they are part of an interdependent organization.

The Factory Building

The construction of Arkwright's textile mill at Cromford in 1772 marks the birth of the factory building. This building type was a response to the functional needs of the mechanization of textile production with its centralization of labor and continuous mechanized production powered at first by water wheel and later by the stationary steam engine.

It was clearly most efficient to concentrate mechanized work in a single site. The hand-loom weaver in his home producing on a small scale had wool delivered to him by an agent, and then had the woven material picked up, often by that same agent for sale. With the vast increase in the scale of production and the availability of the railway, cotton and wool in bulk could now be moved by train to the mill and the finished textiles could be picked up at the mill for distribution on the railway network through England and eventually to the world.

If a small room sufficed for the hand-loom weaver, the factory had to provide many large open areas to hold the many specialized machines with their differing functions, as well as their human tenders. The average mill employed around 150 people. The open space provided for the circulation of workers and the easy movement of materials. The interior space had to be as open as possible to allow the method of continuous production in which the material flowed from process to process with each step in manufacture taking place in a different area. The cotton was washed, then moved to the carding machines that created parallel threads. From there the threads were moved to the spinning rollers, thence to the power looms, and finally to the finishing area. The buildings were constructed with up to five stories with elevators made of cast iron and powered by steam to move material and workers between floors.

The factory building itself embodied the technological innovations of its time. As with the Iron Bridge and the train sheds of the London railway termini, the structural strength of cast iron made possible the spanning of these large interior spaces. The seemingly limitless rooms of the factory became another emblematic space of the machine age. Iron beams spanned the large room. As in the railway stations, the beams were supported by iron columns. As in the train sheds, panes of mass-produced glass provided light during daylight hours. Production continued at night with illumination from coal-gas lamps. Ventilation, however, was wholly inadequate, if not wholly absent. The air was filled with tiny fibers from the cotton and wool being spun and woven. This omnipresent fluff, as it was called, created a serious health hazard for the men, women, and children who worked in the mills, as well as a serious fire danger. With the flammable fibers in the air, the early mills constructed of wood were particularly vulnerable to catastrophic fires. In order to reduce the danger of fire, by the Victorian period mills were built of brick on a cast-iron frame, although the floors were still of wood.

Although the architecture was modern in its use of mass-produced iron and glass for the working interior, as in the case of the train sheds it seemed necessary to the Victorians to decorate the structure in architectural styles of the past. Perhaps this decoration was a means of camouflaging the building's actual use. One of the most popular styles for factories was based on the buildings of Renaissance Italy. But the Victorian towers modeled on those of pre-industrial Italian cities such as Venice had a practical purpose. The high, arcaded towers held bells that rang to summon the nearby workers to their daily shifts.

Power in the Factory

If the domestic weaver could power his loom by his own hands and feet, mechanization in centralized sites called for power of a vastly greater magnitude. The first textile mills, such as the Cromford mill of Arkwright, employed water power. Waterfalls and fast-moving streams turned large water wheels on the exterior of the factory building. This dependence on water power narrowed the choice of siting for mills to locations at streams that had falling water. Thus mills generally located in rural areas or in small towns. The celebrated American mills at Lawrence and Lowell, Massachusetts, that Dickens felt compelled to visit on his American tour took advantage of the fast-moving rivers of New England.[19]

From the 1790s, the steam engine rapidly replaced the external water wheel. The application of the Watt stationary steam engine to spinning machinery and to the power loom proved to be highly efficient. By 1823, there were approximately ten thousand steam-powered looms in operation in Great Britain. By the first decades of the Victorian age, the use of steam to power mills and factories had become virtually universal. The reliability of steam engines and the availability of coal delivered by the railways to power these engines eliminated the need to site factories by fast-flowing streams and allowed their placement anywhere. Thus the application of the steam engine to mechanized manufacture made possible the sudden growth of the Midlands industrial cities.

The entire factory was powered by one or two steam engines. These coal-fired stationary engines were usually situated in a room within the factory or a building within the mill complex separate from the work space. Power was transmitted from the engines to the individual machines, the spinning jenny and the looms, by a complex mechanism of gears and rods. The vertical movement of the engines' pistons was transformed into rotary motion by a set of gears. This gearing then turned a series of shafts that ran through the factory. Individual machines were connected to these rapidly spinning rods by leather belting of varying sizes that regulated power to the needs of the machine. The factory interior was filled with constant rotary and vertical motion of metal and leather, a dangerous place in which one could lose a limb or a life by being caught in a rotating shaft or moving belt.

Work in the Factory

From the time of the building of the first factory by Arkwright at Cromford in 1772, mechanized production was organized according to the principle of the division of labor. The principle was first enunciated in 1776 by the economist Adam Smith, the secular patron saint of Victorian industrialists, in *The Wealth of Nations*. Smith extolled the benefits of splitting the manufacture of a single object into separate repeated activities by taking the example of the "trade of the pin-maker": "One man draws out the wire, another straightens it, a third cuts it, a fourth points it, a fifth grinds it at the top for receiving the head; to make the head requires two or three distinct operations; to put it on, is a peculiar business, to whiten the pins is another ... the important business of making a pin is, in this manner, divided into about eighteen distinct operations." Generalizing from this example, Smith prophetically states: "The division of labour ... so far as it can be introduced, occasions, in every art, a proportionable increase of the productive powers of labour."[20]

Smith turned out to be correct. The principle of the division of labor proved to be startlingly effective when applied on a large scale to mechanized production. No longer did a single weaver produce a single cloth. The process through which the fiber material was transformed to finished cloth was split into separate mechanized operations. Bales of cotton imported from the American South or Egypt arrived at the mill by the railway from major ports such as Liverpool. The fiber was washed, then drawn into parallel strands within the drum of huge carding machines where iron teeth combed the cotton in preparation for spinning. These machines ran automatically, with the only task of the operatives to feed the raw material to the toothed drums, then stop the machine for a moment if snags developed. This carding was hot and dirty work and performed mainly by women in a room filled with cotton fluff.

The carded or combed cotton was then taken to the spinning machines or mules that twisted the fibers into yarn or thread. Spinning was done on the self-acting mule, which only needed a spinner to control the operations of the machine, sweep up debris, and keep an eye on the process in case of mechanical difficulties. One danger to continuous production lay in the breaking of threads during spinning. Specialized piecers were employed to connect the threads when they separated. And since the space within the spinning machinery was small, the dangerous work of tying broken threads within the moving machine was given to smaller people with smaller fingers, that is, to children and to women.

From the spinning machinery, the yarn collected on spindles or bobbins was transported to the looms in the weaving rooms where another set of specialized operatives worked. Workers, both male and female, would string the frames of the loom. Once in operation, the basic action of moving the woof through the warp was done not by hand but by mechanized shuttles that took their energy from the rotating rods above the looms. In these automated looms the shuttle moved ceaselessly back and forth through the web, carrying the woof or thread through the warp, creating the finished pattern. If the factory used a jacquard loom for more detailed patterns, the design had been programmed into the punched paper cards that directed the raising and lowering of threads. Males were usually given the skilled task of operating the power loom itself. Women and children loaded the spindles for the loom, watched over the loom in its operation, and gathered the finished cloth. As in the spinning process, children were most often used to repair broken threads within the moving power loom. The mill employed engineers, to use the Victorian term, to maintain the stationary steam engine outside the factory and the whirling rods that transmitted energy through the building.

Factory Discipline

The factory system also brought into being new forms of social organization. The concentration of production in the mechanized factory not only moved work from the countryside to the city, but also relocated the work-place from the home to a separate, distant location. We should remember that the pre-industrial hand-loom weaver worked in a domestic space in an isolated cottage in a rural area. Home and workplace were the same. With the centralization of work in the urban factory, both male and female laborers left home each day for work, returning at the end of the workday. Although this pattern of the separation of labor and home now seems normal, it is important to remember that the workday spent away from home was new to the nineteenth century, a direct result of the rise of the machine. With the contemporary innovation of the Internet, this split between home and work is gradually dissolving. In any case, nineteenth-century mechanized production on a large scale created what the Victorians described as separate spheres—home and work, domestic and industrial, family and business.

The Victorians associated the domestic sphere with the female and the industrial sphere with the male, because middle-class women did not work. But many working-class women left their homes at some point in their lives to labor in the mills. A high proportion of factory workers were women, in some periods more than 60 percent. Many were married women, often form-ing one-third of the female labor force. Women were heavily concentrated into the preparation of cotton for spinning in the card room and in repairing broken threads in spinning. Children, too, worked in the mills as they had worked within domestic industry and on family farms. Before 1870, there often were no schools available in the new factory cities. And the wages paid to children, though small, boosted the family's earning power.

With the rise of the machine, a social discipline was needed to create a productive workforce suited psychologically to work with the machine. Workers who had come from the countryside to the industrial city had to be converted from the internalized rhythms of rural life to accept the rhythms of the machine and of the factory. Mechanization of work meant the mechanization of time itself. The length of the farming day was deter-mined by the sun. Farming tasks depended on the cycles of nature—a fallow winter for rest, intense work at planting and harvest time. Hunting depended on the patterns of animal life. In the factory, the machines ran regardless of season. Water wheels might have to stop for drought or flood-ing, but the steam engine carried on tirelessly and could be speeded up or continued for extra time by the management. In a mode of regulation of the self that has become so commonplace that it is perhaps difficult to feel its newness with the emergence of machine production, one worked not accord-ing to the rhythms of nature, but rather to a set number of hours. The du-ration of work and the respites from work were governed by the central machine of the industrial age, the mechanical clock invented in the medieval monasteries to regularize the routine of prayer.[21]

Just as the railway system called for a synchronized and standardized railway time, so the factory system demanded what can be called factory time.
Only through the regulation of clock time could the employer guarantee a
set number of people at a set time in the factory and keep this workforce
working for a set period at a set number of mechanical tasks. The clock
tower of the factory called out the time for beginning and ending work. In
the summer, work began at daylight, in winter in the dark, but always
at the same hour on the clock. Urban industrial workers, male and female,
left their cramped homes to walk to the mill early in the morning, where
their arrival was registered. A system of punishments, usually docking of
pay, compelled the employees to arrive "on time"; workers were fined for
late arrival. And because many factories ran on round-the-clock shifts, the
times of beginning and ending work for each shift had to be carefully coordinated by the clock.

The hours of labor set by the clock were long for men, women, and children. Before the government reforms that gradually took effect during the
Victorian age, adults commonly worked fourteen hours a day, and children
more than nine. During these long hours in the factory, each worker was
assigned a set place from which he or she could not move. Since the
machine did not rest, neither could the workers. Undivided, continuous
attention was demanded. Children were sometimes subjected to beatings for
falling asleep over their work. No breaks, no conversations, no wandering
around the factory floor. In contrast to the relatively free working conditions of the hand-loom weaver in his own house, once in the factory the
workers were constantly watched by supervisors who noted slackness, any
turning from work, any infraction of rules, any lack of concentration on the
job. Thus, the factory system depended upon a form of constant surveillance
by monitors over the body of the individual and also of his or her inner life
so as to create a constant concentration. The personal need for such attention was intensified since workers were often paid by a piecework method
based on the amount of material produced by the machine under their care.

In the factory there was certainly drudgery and boredom. But the flow of
mental energy continued even if the body was controlled. These were not
wholly inhuman places. People could get to know each other well; they provided mutual support; and they engaged in the wonderfully human pastime
of gossip. Women spent the workday with people outside their own family
circle and, in sharp contrast to rural life, with men who were strangers.
From all contemporary accounts, being freed from the rigorous traditional
constraints of village life and mixing with people outside the family led to a
certain erotic edge among factory hands. The social and psychological disciplining of workers within the mechanized factory then, as now, was defeated
by the workers themselves.

2

The Living Machine and the
Victorian Computer

In the nineteenth century, technological innovation brought into being a new kind of machine, a machine that ran without the need of human intervention. With the development of feedback systems such as the steam-engine governor, engines became fully self-regulating, in Victorian terms, self-acting. In the textile mills, the jacquard loom automatically wove complex patterns by carrying out programs punched into paper cards. Expanding the automated mode of the jacquard loom, Charles Babbage developed what he saw as an engine that could think. His Analytical Engine stands as the forerunner of our own programmed computers. With the invention of machines that installed the heretofore distinctly human qualities of autonomy and intelligence into material forms of iron and wood, the traditional distinction between the living and the mechanical, the human and the machine appeared to dissolve. In the Victorian period, as in our own time, with the rise of the intelligent machine human beings came to worry that they were little different from the complex machines of their age. The Victorians began to imagine themselves as intricate self-regulating mechanisms and as energy-generating engines. This chapter will examine how the concept of the living machine informs technological innovation in the Victorian period.

THE LIVING MACHINE

Dorothy Wordsworth's journal of her 1803 tour of Scotland with her brother William and his friend Samuel Taylor Coleridge, both to become celebrated Romantic poets, describes her first contact with a steam-powered pumping engine:

> When we drew nearer we saw, coming out of the side of the building, a large machine or lever, in appearance like a great forge-hammer, as we supposed for raising water out of the mines. It heaved upwards once in half a minute with a slow motion, and seemed to rest to take breath at the bottom, its motion being accompanied with a sound between a groan and "jike." There would have been something in this object very striking in any place, as it was impossible not to invest the machine with some faculty of intellect; it seemed to have made the first step from brute matter to life and purpose, showing its progress by great power. William made a remark to this effect, and Coleridge observed that it was like a giant with one idea.[1]

In this first sight of the stationary steam engine, to Dorothy and William Wordsworth, as to Coleridge, the machine seems to be alive. The engine inhales, "seemed ... to take breath." It even speaks "with a sound between a groan and 'jike.'" The new machinery seems to manifest the free will that depends upon mind. Coleridge nicely describes this sense of physical power combined with a virtually human self-determination in a simile, "it was like a giant with one idea." Dorothy Wordsworth notes that the engine "seemed to have made the first step from brute matter to life and purpose." Life and purpose implies intelligence: "it was impossible not to invest the machine with some faculty of intellect."

The sense that machines were somehow alive grew through the nineteenth century, strengthened by innovations in automatic machinery, especially the development of feedback mechanisms. Andrew Ure, the chief apologist of mechanized production in the early nineteenth century, saw these self-acting machines as giving a form of life to inanimate matter. He writes in *The Philosophy of Manufactures*, in 1835: "The philosophy of manufactures is therefore an exposition of the general principles on which productive industry should be conducted by self-acting machines.... It is in a cotton-mill ... that the perfection of automatic industry is to be seen; it is there that the elemental powers have been made to animate millions of complex organs, infusing into forms of wood, iron, and brass an intelligent agency."[2]

Victorian automatic machinery, then, appeared to manifest the self-regulation and intelligence of a human body endowed with soul. In the mills, the jacquard loom wove complex designs according to patterns programmed into punch cards. The steam-engine governor in the stationary engine and the locomotive regulated the enormous but variable heat energy of the engine without human intervention. It is but a short step from the self-acting jacquard loom and steam engine to the Victorian thinking machine.

THE INTELLIGENT MACHINE: THE VICTORIAN COMPUTER

By the early nineteenth century, there was a widespread anticipation of bringing "metal close to rationality"[3] by expanding the self-regulating machine from the production of patterned cloth to the production of thought itself. This task of inventing a machine that thinks was taken up by Charles Babbage and to a certain extent achieved in his proto-computers, the Difference Engine and the Analytical Engine.

A professor of mathematics at Cambridge University, Babbage moved restlessly from the realm of pure mathematics into the emerging world of machine technology. In an age that mechanized traditional processes like weaving and spinning, Babbage sought to mechanize mental processes, in his words "to calculate by steam."[4] His project was motivated by the need in the age of empire for error-free mathematical tables essential for navigation at sea. As Babbage tells the tale, he saw in the methods of a French mathematician in producing these tables the application to mental labor of the industrial principle of the division of physical labor: "It has been shown, that the division of labour is no less applicable to mental productions than to those in which material bodies are concerned."[5] In France, the work of compiling the tables was divided into three sections depending on the mathematical skills required. In this system, the people performing the routine arithmetical calculation were called "computers." The shift in the term from persons to engines encapsulates the development of the intelligent machine. Babbage felt with some qualification that not all mathematical skills could be reproduced by machinery, but that "part executed by the third class [the computers] ... may almost be termed mechanical."[6] Note here that Babbage is leaping to the notion that the mental activity of arithmetical calculation can be described in the same terms as the actions of physical machinery and thus could be embodied in a machine that he imagined as a "calculating engine."[7]

In 1822, Babbage completed the prototype of a machine called the Difference Engine that would automatically produce error-free mathematical tables. The device, which was never finished, worked according to the mathematical method of finite differences that employs repeated addition as a way of doing multiplication and division, hence the term difference. The term engine suggests Babbage's sense of the similarity of this machine to the automatic machinery of the time. In planning the Difference Engine, Babbage followed the main principle of nineteenth-century technological change in replacing human operations by mechanical in a complex self-regulating device. It must be stressed that the significance of Babbage's calculating engine lies in that it is not physical but mental activity that is done by the machine. It is the first known device to "embody mathematical rule in mechanism."[8]

The Difference Engine consisted of an elaborate clockwork mechanism within a cast-iron frame. Using the decimal system, numbers are represented on geared brass wheels called figure wheels. Each wheel is turned to a specific gearing to represent a number, e.g., a wheel will be turned two teeth to represent the number 2. The wheels are arranged on columns centered on brass rods with the place of the wheels on the column representing their place in the decimal system. For example, for the number 246, the wheel turned to six for units would be at the bottom, with the figure wheel for four in tens above it, and at the top the wheel turned to two for hundreds. For his Difference Engine, Babbage imagined figure wheels piled for as many as twelve digits on a single column and a total of six adjoining columns.

Before operation began, all the figure wheels were set at a specific starting value. Then the Engine was set in motion by an operative turning a large hand crank: "The figure wheels would be driven in their deliberate dance by trains of gears and levers derived from this single source."[9] The columns would rotate and the figure wheels mesh with figure wheels representing numbers on adjacent columns. A system of levers would continuously add by carrying numbers on the figure wheels to another adjacent column until a final figure was reached.[10] The mechanism thus transformed physical force into what had hitherto been mental power.

Like the steam-engine governor and the jacquard loom, the Difference Engine is fully automatic. The class of people known as computers used rudimentary calculating devices that demanded constant manual manipulation with the accompanying possibility of error. But once set in motion, the Difference Engine adds and multiplies without human interference and thus without human error. Babbage asks, "whether the numbers on which it is to operate are placed in the instrument, is it capable of arriving at its result by the mere motion of a spring, descending weight, or any other constant force? If the answer be in the affirmative, the machine is really automatic; if otherwise, it is not self-acting."[11] The Difference Engine anticipates Babbage's Analytical Engine as well as the computers of our own time in following a set, if limited, program. Here the operations are limited to calculation and initiated by setting the figure wheels at a particular value.

Building the Difference Engine presented enormous technical problems. The many figure wheels with their gearing required precise machining to be exactly alike in their dimensions in order to mesh with other figure wheels in the process of carrying sums to another column. To create the hundreds of precise parts needed, Babbage collaborated with a highly skilled machinist and draftsman, Joseph Clement. Clement drafted the schematics of the parts from Babbage's sketches, then oversaw the precise machining of the brass

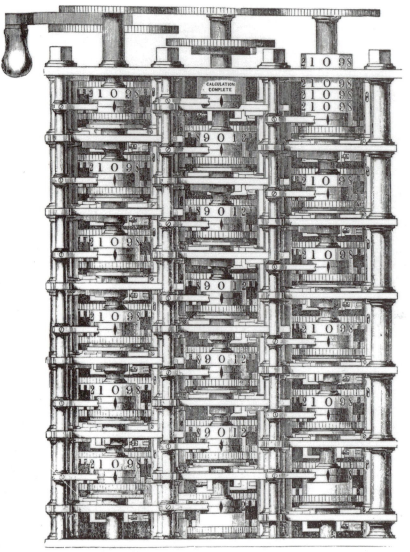

B. H. Babbage del.

Charles Babbage's Difference Engine. The sketch shows the geared brass wheels Babbage called "figure wheels." Each wheel can be turned to a specific gearing to represent a decimal number from 0 to nine marked on each toothed wheel. The wheels are arranged on columns centered on brass rods with the place of the wheels on the column representing their place in the decimal system. To operate, the wheels are turned by the crank at the top. Impression from a woodcut by Benjamin Herschel Babbage, Charles Babbage's son. [Reprinted with permission of Science Museum/Science & Society Picture Library.]

wheels, columns, and levers. This collaboration of theoretician and artisan indicates, again, how necessary was the collaboration of craftsman and theoretician in Victorian innovation. The building of the Difference Engine, like the building of the early steam engines, called for innovations in engineering to achieve exacting tolerances. And, as in the development of the steam engine in the artisanal workrooms of James Watt, such technological innovation depended on skilled craftsmen to give ideas material form.

The building of the Difference Engine, however, never progressed beyond the prototype stage. A "finished portion of the unfinished engine" was delivered to Babbage in 1832, but a complete Engine was never built by him.[12] The reasons are various. A completed Engine would have needed twenty-five thousand precisely machined parts, a number that would have demanded an enormous expenditure of time and labor. And since the parts needed to be machined to narrow tolerances, without the exacting standardization of machine tools throughout England such parts could not be created by separate suppliers. Babbage had to depend on the work of one supplier and one machine, that of Clement. Overwhelmed by the task, Clement quit his work in 1832.

There were financial difficulties, too. The British government had in 1823 provided a large grant for development work in the hope of obtaining wholly reliable astronomical tables needed for the Royal Navy. But as Babbage demanded more and more money, the government provided only small and insufficient amounts, in acrimonious exchanges with Babbage. In 1842, the government finally refused to provide additional funding. Thus the Difference Engine was imagined but never built in the Victorian age.

There remained a question, whether the Difference Engine could have been built and could have functioned in the nineteenth century had money been available. To answer the question, the Science Museum of London in 1991 decided to build a full-scale Difference Engine using only Babbage's designs and the techniques and materials available to him in the nineteenth century. The museum built a Difference Engine within Victorian limitations, and the machine worked. The completed Engine measures seven feet high, eleven feet long, and eighteen inches deep. It holds four thousand parts. It operates cleanly and calculates automatically and accurately, just as Babbage said it would.[13]

For all the brilliance of its conception, the Difference Engine was limited in its function. It was essentially a fully automatic, error-free calculating machine. Yet after working on the Difference Engine, Babbage became convinced that a machine could imitate the full range of human mental life, particularly in the realm of logical analysis. He was certain that "the whole of the developments and operations of analysis are now capable of being

For an intelligent alternative history in which the Difference Engine had been built in the nineteenth century and Victorian England became a computerized society, I recommend William Gibson and Bruce Sterling's cyberpunk novel, *The Difference Engine*. In this alternative nineteenth century, Difference Engines are employed to control the society. At the center of government are "giant identical Engines, clock-like constructions of intricately interlocking brass, big as rail-cars set on end, each on its foot-thick padded blocks. The white-washed ceiling ... was alive with spinning pulley-belts, the lesser gears drawing power from tremendous spoked fly-wheels on socketed iron columns."[14]

executed by machinery."[15] To this end, Babbage moved beyond his conception of the Difference Engine to plan what he saw as a full thinking machine. He called this advanced machine the Analytical Engine.

Like the computers of our own time, the Analytical Engine was designed to consist of two parts. As described by Babbage, these are

1. The store in which all the variables to be operated upon, as well as all those quantities which have arisen from the result of other operations, are placed.
2. The mill into which the qualities about to be operated upon are always brought.[16]

Babbage imagined the "store" as a device where numbers were to be kept on geared figure wheels attached to columns as in the Difference Engine. These numbers were to be transported to the "mill" for processing by long-toothed racks. In our terms, the store would correspond to the memory of the modern computer. The mill is the place where the numbers are processed. The physical separation of memory and central processor, of store and mill, is the basis of modern computer construction.

Babbage's terminology and methods for his Analytical Engine illustrate how his thinking machine emerged from the automatic machinery of the Victorian factory. The store corresponds to the term for the storehouse of raw material in the Victorian textile mill. In this computer memory, the numbers made material in the figure wheels correspond to the bales of cotton to be moved by cart to the spinning jenny and power loom for fabrication. He uses the term mill for the place where the raw material of numbers is processed. His terminology suggests that the Analytical Engine was to him a miniature factory where self-acting machinery replicated mental processes to manufacture products of intellect rather than cloth. In remarkably prescient thinking, Babbage imagined the output of the Analytical Engine to be set in material form in a variety of ways: printed automatically on paper as in our contemporary printers or punched into templates for the perforated cards that could be used for further programming.

Babbage's contribution to computing lay in his ideas for instructing the interaction of the store and the mill, of memory and processing. The punched paper cards of the Analytical Engine were much like the cards eventually employed in mid-twentieth-century IBM computers.[17] This storage of data on cards rather than within the machine itself as in the Difference Engine provides a form of what we would call external memory. Since the number of the punched cards did not depend on the physical capacity of the Engine, the amount of memory and the number of instructions was in theory limitless.

By using punched cards, in a quite tangible way Babbage applied the techniques of the textile mill to mental labor; he sought to use the methods of the jacquard loom to set intellectual operations into material form. In this automatic weaving device, textile patterns were encoded into punched holes in paper cards that directed wires to raise different warp threads in an ongoing, complex weaving operation. Thus, the jacquard loom could weave an infinite number of patterns through algorithms programmed into these cards that constituted a form of external memory. Babbage was inspired by this programming function of the loom: "The manufacturer might use the same cards, and put into the warp threads of any other colour. Every thread might even be of a different colour, or of a different shade of colour; but in all these cases the *form* of the pattern will be precisely the same—the colours only will differ."[18]

For the instructions that would initiate the automatic functioning of the Analytical Engine, Babbage employed several kinds of punched pasteboard cards. Operation Cards would tell the mill what kind of arithmetical function (addition, subtraction, multiplication, division) to perform in a set sequence. Thus a function could be repeated a limitless number of times. Variable Cards would specify from where in the store the information was to be taken. Number Cards would set the geared wheel of the store in sequence. There was even a plan for conditional branching, the "if ... then" sequence in the programming that instructs that if a certain result occurs, then another sequence is to be followed. Also, in both the loom and the Engine the "presence or absence of a hole in a punched card corresponds to 'on' or 'off,' or in binary notations, to zero or one."[19] Thus the binary system which is the mathematical principle at the center of modern computing lies also at the heart of the jacquard loom and the Analytical Engine.

After years of working in private, Babbage announced his plans for the Analytical Engine publicly in 1840. He found an unexpected champion for his proposed engine in Ada, Countess of Lovelace, daughter of the poet Lord Byron. Lovelace was that most rare nineteenth-century figure, a female mathematician. Her writings publicizing the Analytical Engine illuminate Babbage's thinking and also demonstrate that the nineteenth century realized that an unprecedented machine, a thinking machine, was emerging: "We are not aware of its being on record that anything partaking in the nature of what is so well designated the Analytical Engine has been hitherto proposed, or even thought of, as a practical possibility, any more than the idea of a thinking or of a reasoning machine."[20] Lovelace emphasizes how the Analytical Engine extends the innovations of automatic textile machinery. She notes "how the machine can of itself, and without having recourse to the hand of man, assume the successive dispositions suited to the operations" since the "solution of this problem has been taken from Jacquard's apparatus ... The analogy of the Analytic Engine with this well-known process [of the Jacquard] is nearly perfect."[21] And in a brilliant comparison that sums up the interconnected technological innovations of the age, she neatly states, "The Analytical Engine weaves algebraic patterns just as the Jacquard loom weaves flowers and leaves."[22]

As public spokesperson for the Analytical Engine, Lovelace presents the Analytical Engine as a true thinking machine. She assumes that the work of the intellect itself is analogous to that of a physical machine since both operate according to a set of laws or algorithms. Using the futuristic narrative that governs so much discussion of intelligent machines in the nineteenth century, Lovelace sees in the Victorian fusion of the mental and the mechanical an anticipation of the computers of our own age: "The imagination is at first astounded at the idea of such an undertaking [a machine capable of analysis]; but the more calm reflection we bestow on it, the less impossible does success appear, and it is felt that it may depend on the discovery of some principle so general, that if applied to machinery the latter may be capable of mechanically translating the operations which may be indicated to it by algebraical notation."[23]

Like Lovelace's imagined computer of the future and the Difference Engine, the Analytical Engine was never constructed. Although parts of the mill were built, discouraged by his continuing inability to construct a full Difference Engine, Babbage never made a concerted effort to put the Analytical Engine into completed material form.

Babbage's effort to mechanize thought in his Engines was celebrated as exemplifying the technological progress of the age. Important visitors, including Charles Dickens and Prince Albert, husband of Queen Victoria, flocked to Babbage's house to admire the completed parts of the Difference Engine. In 1834, Dionysus Lardner, the science correspondent of *The*

Edinburgh Review, wrote, "A proposition to reduce arithmetic to the dominion of mechanism—to substitute an automaton for a composition—to throw the power of thought into wheel-works could not fail to capture the attention of the world."[24] On presenting the gold medal of the Astronomical Society to Babbage in 1825, the president of the society lauded the Difference Engine: "Just as the artisan has been furnished with commands of power beyond human strength, joined with precision surpassing any ordinary attainment of dexterity, [the calculating engine] substitutes mechanical performance for an intellectual process; and that performance is effected with celerity and exactness unattainable in ordinary methods."[25] Yet it must be noted that for all the anticipations of the modern computer in the Analytical Engine, the actual history of the electronic computer shows that its inventors did not look back on upon Babbage's plans, which became known only after the innovative work on the modern computer in the mid-twentieth century.

VICTORIAN SCIENCE FICTION

Both Lovelace and Babbage imagine the Victorian thinking machine as a newly developed form of intelligent life. Lovelace sees the Analytical Engine as a "mechanism" that can "combine together general symbols, in successions of unlimited variety and extent, a uniting link ... between the operations of matter and the abstract mental processes of the most abstract branch of mathematical science."[26] Babbage felt that working with the Analytical Engine would be communicating with an entity endowed with mind. He speaks of "the real difficulty [in] teaching the engine to know when to change from one set of cards to another."[27] He even sets an improved thinking machine in a science-fiction narrative of an intelligent machine that wields an imperious power over their human attendants. In the days before machines could be imagined as capable of speech, the engine commands by ringing a bell:

> I then added that when the machine wanted a tabular number, say the logarithm of a given number, that it would ring a bell and then stop itself. On this, the attendant would look at a certain part of the machine, and find that it wanted the logarithm of a given number, say of 2303. The attendant would then go to the drawing containing the pasteboard cards representing its table of logarithms. From among these he would take the required logarithmic card, and place it in the machine. Upon this the engine would first ascertain whether the assistant had or had not given him the correct logarithm of the number; if so, it would use it and continue its work. But if the engine found

the attendant had given him a wrong logarithm, it would then ring a louder bell and stop itself. On the attendant again examining the engine, he would observe the words, "Wrong tabular number," and then discover that he really had given the wrong logarithm, and of course he would have to replace it with the right one.[28]

Thus, it is in the nineteenth century that there emerges the figure that haunts our own science-fictional imagination, the machine with a will of its own. The contemporary imagination is even more occupied with the threat and the promise of intelligent machines. In the film *2001: A Space Odyssey*, the supercomputer named HAL attempts to take over the starship and must be unplugged. In *Blade Runner*, the replicants seek revenge against their human creator who has programmed them for death.

In a figure that launched a thousand sci-fi films, Andrew Ure, the Victorian celebrator of industrialism, imagines the intelligent mind of the human being transferred into the iron machine to form a robot that grows and learns. Fastening on the new self-acting mule developed by Richard Roberts, Ure conjures up "a machine apparently instinct with the thought, feeling, and tact of the experienced workman—which even in its infancy displayed a new principle of regulation, ready in its mature state to fulfill the functions of a finished spinner."[29] Anticipating our own science-fiction terminology, Ure calls such living machines Androides: "[machines] of this kind, are sometimes styled Androides, from the Greek term, like a man."[30] For Ure and for most Victorians, these Androides, machines that take on human functions, are not monstrous, but beneficent, at least according to the industrial ethics of the time. Ure notes such living machines have freed industry from the need to employ children: "The self-actor has already dispensed with the services of many young children, and in its further progress will dispense with many more."[31]

In a phrase of Ure's often quoted by the Victorians, the spinning mule as Androide becomes a person: "SELF-ACTOR MULE, or the IRON MAN, as it has been called in Lancashire."[32] In the illustration in Ure's *Dictionary of Arts* of the mule, the device visually becomes an iron man or metal person. Quite intensely, the drawing suggests the organic qualities of self-acting devices of metal. We see in the Androide the suggestion of a head, extended arm, even lines of vision from the head to the spinning mechanism. Similarly, the self-acting drawing frame for spinning cotton becomes in Ure's mind a mechanical replication of a human action: "An exact imitation of what takes place when we draw a tuft of cotton wool between our fingers and thumb in order to ascertain the length of the staple."[33]

Ure looks beyond the Iron Man to its creator: "The Iron Man, as the operatives fitly call it, sprung out of the hands of our modern Prometheus [Roberts] at the bidding of Minerva—a creation destined to restore order among the industrious classes, and to confirm Great Britain in the empire of art."[34] The naming of Roberts, the inventor of the Iron Man, as a "modern Prometheus" provides a self-conscious contrast to Victor Frankenstein, the creator of what is generally termed the Frankenstein monster. The full title of Mary Shelley's book is *Frankenstein: The Modern Prometheus*. Ure proposes a myth that runs counter to Shelley and our own easy labeling of new inventions as "Frankenstein monsters" to emphasize the destructive power of new technologies. Instead, nineteenth-century inventors are seen by Ure and many Victorians as repeating the work of the ancient human being Prometheus who stole fire from the gods to the great benefit of mankind.

BODY AS MECHANISM

By the mid-nineteenth century, then, Britain appeared to be populated by machines that could act independently, regulate themselves, calculate, and even, it seemed, think. From steam engines to Babbage Engines, machinery imitated the physical actions and mental processes of human beings. The wires of the jacquard loom replaced the fingers of the weaver. The figure wheels and columns of the Difference Engine mimicked the calculating powers of the human mind. As Andrew Ure noted in the 1830s, the self-acting machine is "that class of mechanical artifices in which the purposely concealed power is made to imitate the arbitrary or voluntary motions of living beings."[35] And given this seeming interchangeability of the organic and the mechanical, it is the mechanical that is often superior. The jacquard loom worked tirelessly and without error, as did the Difference Engine.

For the Victorians, the rise of the self-acting, intelligent machine seemed to collapse the traditional distinction between the human and the mechanical. If cast-iron machines might be seen as living beings, so human beings might be seen as living machines. This conception of the living body as machine, a set of ideas termed mechanism, had been articulated in its most influential form by the French philosopher René Descartes in the seventeenth century. Descartes imagined that organic bodies are subject to the same laws of physics that govern physical objects. The bodily machine can be explained on the same mechanistic principles as, for example, the geared works of clocks powered by springs or the levers and pulleys that lift heavy loads. The Victorians often employed this simple equation of machine and body to analyze contemporary work. Peter Gaskell, a critic of industrialism, speaks in 1836 of the work of the hand-loom weaver in terms of muscles transmitting force through a mechanical system of levers: "The position in which the weaver sits is not the best for muscular exertion, as he has no firm support for his feet.... He has thus to depend for a fulcrum chiefly on the muscles of his back, which are kept in constant and vigorous action, while one order of muscles is employed with little power of variation, in moving the shuttle and beam."[36]

For Descartes, only the lower animals are purely mechanical devices. Human beings differ in possessing a soul that is not material and thus not subject to physical law. This idea is often termed the ghost in the machine. But this religious notion was clearly challenged by the Victorian rise of the machine. The self-acting intelligent machines of the nineteenth century seemed a refutation of the idea that human action could only be explained by positing a soul. Victorian automatic machines such as the steam engine regulated themselves through feedback, not an immaterial soul. The Babbage Engines were designed to manifest intelligence without a ghost. For Babbage and Lovelace, as for builders of artificial intelligence in our own time, mental processes follow natural law as much as do the movements of the physical body.

The Victorian sense of the body as machine also emerges in the nineteenth-century occupation with what we now call cyborgs, that is, entities in which the mechanical and the organic are inextricably spliced.[37]

The vision of man fused with machine is due in part to the increased nineteenth-century use of prosthesis, a material device that substitutes for a body part as does a wooden leg or an iron hook as hand. Artificial limbs were needed more and more to replace body parts torn off by the unfenced machinery of the factory. Ure writes in a rather unfeeling if utilitarian fashion of the fusion of the inanimate mechanical device with the body: "the substitution of tools for human hands ... [being] used to assist the labour of those who are deprived by nature, or by accident, of some of the limbs [in workmen] enabled them to execute their tasks with precision, although labouring under the disadvantages of the loss of an arm or leg."[38]

On a larger scale, for the Victorians the power loom or steam hammer seemed to become spliced with the operator so as to abolish the distinction between machine and human. The celebration of this fusion is exemplified in the autobiography of James Nasmyth, a Victorian famous in his time for his invention of the steam hammer, a device that directed the energy of steam to such industrial tasks as beating cast iron into girders or driving pilings into the ground for bridges and viaducts. For Nasmyth, the operator of the steam hammer becomes, in a modern version of Descartes's theory, the internal mind of the iron body: "Thus, by the more or less rapid manner in which the attendant allowed the steam to enter or escape from the cylinder, any required number or any intensity of blows could be delivered. Their succession might be modified in an instant; the hammer might be arrested and suspended according to the requirements of the work. The workman might thus, as it were, *think in blows*."[39] Nasmyth's own drawing of his steam pile driver exhibits the science-fictional imagination of the day. Here within a small compartment is the human as a brain dwarfed by and yet controlling the hard, elongated metal body part. As a cyborgian entity of brain within a mighty metal body, the giant pile driver becomes a giant prosthesis of masculine power, a kind of Victorian Robo-Cop.

BODY AS ENGINE

As the invention of automatic intelligent machinery challenged the notion of the human as a ghost in the machine, so innovations in the steam engine brought another redefinition of the human body. The nineteenth-century development of increasingly efficient steam engines went hand-in-hand with the growth of the new science of thermodynamics, the branch of physics that deals with the relationships between heat and other forms of energy. For the Victorians, this interrelated practice and theory led to the radical reconceptualization of the active human as essentially an engine: "The working body was but an exemplar of that universal process by which energy was converted into mechanical work, a variant of the great engines and dynamos spawned by the industrial age.... For European physicists and physiologists, Descartes' distinction between the animal machine and the human being was no longer meaningful.... The automata no longer had to be denied a soul—all of nature exhibited the same protean qualities as the machine."[40]

This model of the body-as-engine had deep-seated implications. For one, the notion undermines the traditional idea that a non-material soul implanted by God energizes the machine of the human body. Instead, this religious idea is supplanted by the mechanistic theory that energy in the body, in a way consistent with the laws of thermodynamics, is generated by the combustion of food as fuel. Quite simply, "the human body and the industrial machine were both motors that converted energy into mechanical work."[41] The fuel, whether coal or food, is burned and thereby transformed into heat energy within the boiler or within the interior of the body. Within thermodynamic terms, this energy is converted to work defined as movement, whether of pistons or of muscles.

Even before the advent of thermodynamic theory, the mechanists of the early industrial age equated their cast-iron coal-burning engines with the animals that the engines replaced. Commonly, the energy requirement of the engine was compared to the organic image of ingesting food as a fuel that is transformed into energy. In a nice example of the industrial universe asserting its continuity with the old agricultural order, in 1783, James Watt had standardized the unit of energy produced by his steam engines in terms of horsepower as a way of selling his engines to those unfamiliar with the new machines.[42] Commonly in the nineteenth century, the comparison of the steam engine as consuming coal to the horse ingesting food and water became a way of comprehending and celebrating the new mechanical source of energy. For Edward Baines, "All [textile machines] derive their motion from the mighty engine, which, firmly seated in the lower part of the

building, and constantly fed with water and fuel, toils through the day with
the strength of perhaps a hundred horses."[43] The metric of horsepower is
still used to measure the power of engines in the age of the automobile.

Reconceptualizing the human being as engine also transformed the issues
of discipline within the mechanized factory. In the early industrial age, the
problem in managing the first generation of industrial workers was the difficulty of habituating men and women accustomed to the changing seasonal
rhythms of agricultural life to clock time and constant attention to their
machines. With the human body seen as an engine, the main problem for
productivity became managing the energy generated by the human motor.
Implicit in the thermodynamic model of human labor is the threat that
exhausting mental and physical work over long hours may so deplete the
energy of the worker that he or she can no longer function effectively.
Thus, in the later nineteenth century the chief issue for efficiency in industrial work became adjusting the relations of human and iron engines in
regard to the discharge of energy. The aim of industrial management
became producing at maximum efficiency while avoiding the exhaustion of
energy that we call fatigue. The crucial sign of mismanagement became not
the idleness that the early Victorian factory owners feared, but the mental
and physical fatigue that comes from inefficient regulation of the human
motor.

For Charles Babbage the question of fatigue is key to any analysis of
mechanized manufacturing. His thinking moves to the question of fatigue
in terms of the physiology of the conservation and expenditure of energy in
the muscles. He notes, for example, the need for proper management of
energy expenditure in order to achieve maximum efficiency: "The fatigue
produced on the muscles of the human frame does not altogether depend on
the actual force employed in each effort, but partly on the frequency with
which it is exerted.... It does therefore happen, that operations requiring
very trifling force, if frequently repeated, will tire more effectually than
more laborious work. There is also a degree of rapidity beyond which the
action of the muscles cannot be pressed."[44]

For the mechanists, the mind was also conceived as an engine that produces energy. Thus, there emerges the need to regulate mental energy so as
to prevent the mental fatigue that appears in factory workers in a range of
debilitating emotional symptoms. Peter Gaskell writes in 1836 that for child
laborers "the perpetual necessity for attention prevents any thing like bodily repose, so that, although no labour, in the common meaning of the word,
is undergone, fatigue necessarily results."[45] For adults, the tiring effects of
factory labor are similar: "The only fatigue suffered is that brought on by
the necessity for continued attention—a fatigue as injurious to health as
that induced by physical exertion, if not excessive."[46] And "the muscular
system becomes languid, and a degree of irritability is produced, which

exhausts the powers of the nervous system, and excites feelings of great discomfort and depression."[47]

Certainly, our own science of mechanized work depends on the nineteenth-century vision of the human worker as energy-producing motor. Like the Victorian writers on factory labor, our efficiency experts look to managing the rhythm and expenditure of human energy to avoid the debilitating mental and physical fatigue that would slow down mechanized production.

3

The Great Exhibition of 1851:
The Architecture of the Machine
and the Rise of Commodity Culture

No single event so epitomizes the Victorian celebration of the machine as does the Great Exhibition of the Works of Industry of All Nations that opened in the center of London in 1851. Housed within a vast hall of prefabricated glass and cast iron known as the Crystal Palace, this was the first world's fair. The Great Exhibition put on display not only the machines of England at work, but also the cornucopia of machine-made commodities now increasingly available to all economic classes. To this spectacle-of-the-machine show flocked in great numbers the population of the first industrial nation that had also become the first consumer society.

The idea for the Great Exhibition came in large part from Prince Albert, Queen Victoria's German-born husband. His aim was to encourage manufacturing in England, raise the standard of design for British machine-made goods, and demonstrate the position of England as the world's leading industrialized nation. For these purposes, he obtained the cooperation of England's hereditary aristocracy and newly powerful industrialists as well as the encouragement but not the financial support of the government.

The title, The Great Exhibition of the Works of Industry of All Nations, especially the use of the word industry, suggests the complex transformation of England that was celebrated by the Exhibition. The term industry had originally applied to the act of working with diligence and skill, as we now might describe an enterprising person as industrious. It was only in the early nineteenth century that the term industry applied less to a human quality than to the set of institutions organized for the purpose of mechanical production. Thus, by the middle of the nineteenth century, the moment of the Great Exhibition, the term industry had come to refer to what we now call industrialism, the system of machine production.[1] Yet the echo of the earlier meaning remains. Mechanized industry depends on industriousness; mechanized production depends on dutiful and disciplined action by the industrial worker. Furthermore, the term works applies both to the processes and places of industrialism, as in the current term steelworks, as well as to the products of such activity, especially to machine-made goods.

The Great Exhibition was to show the "Works of Industry of All Nations." As the first modern world's fair, it manifested the global perspective of the nineteenth century as machine technology spread to many nations. The Crystal Palace displayed machinery and machine-made products from a number of countries, particularly from the rising industrial nations such as France, Germany, and the United States. And yet the emphasis of the Great Exhibition fell on England as the first and still the preeminent industrial nation. A manifestation of nationalism, the Crystal Palace aimed to demonstrate to the world the technological innovations that were creating the wealth of England.

The Crystal Palace, then, functioned as a temple devoted to the belief that England's technological progress manifested the Will of God. In the movement toward the divinely ordained goal of an industrialized world, England was leading the march of nations. As the future Poet Laureate Alfred Tennyson said in his 1837 poem, "Locksley Hall," "Better fifty years of Europe than a cycle of Cathay."[2] The words articulate the Victorian vision of England and Europe moving forward along a linear track, similar to a railway line, laid out by God toward an industrial future. Meanwhile, non-European nations, such as China (Cathay), remained stuck in cyclical time as they endlessly repeated their age-old craft methods.

This Victorian identification of mechanization as the manifestation of God's Will and of Britain's leadership as a sign of being a nation chosen by God is no better expressed than by Queen Victoria herself on visiting the Crystal Palace:

> Every time one visits this great work, one feels more and more impressed with its lofty conception. The mottos of the Catalogue, chosen by my beloved Albert [Prince Albert, her husband], are very appropriate and significant of the great undertaking, such as "The Earth is the Lord's and all that is therein".... "The germs of every art are implanted within us, and God, our instructor, without our knowing it, develops the faculties of invention".... "The progress of the human race resulting from the labour of all men ought to be the final object of the exertion of each individual. In promoting this end we are carrying out the will of the Great and Blessed God." This latter motto is Albert's own.[3]

ARCHITECTURE OF THE MACHINE

To display the machinery and the manufactures of England and of the world, a vast exhibition space was needed. No existing building was adequate. The Commissioners of the enterprise decided to build for the Great Exhibition a structure that could be erected quickly and, with Victorian prudence, cheaply. It was also necessary that the building could be removed once the Exhibition was over. A site was selected in central London at Hyde Park, a vast open space with a location similar to New York's Central Park. The structure eventually called the Crystal Palace deserves the term often used by the Victorians to describe it: miraculous. It was truly an emblem of its age. As a building the Crystal Palace brilliantly created an architecture of the machine in bringing together the elements of the machine age: standardization of parts; mass production of reliable materials; and the replacement of artisanal labor with the efficient division of labor.

If the vast building that so quickly arose in Hyde Park looked like a giant glass greenhouse, that is exactly what it was. For the design of this building, the Commissioners sponsored a nationwide competition. The winning plan was submitted by Joseph Paxton, the chief gardener for the Duke of Devonshire. Paxton's plan is modeled on the large greenhouses he had built on the Duke's estates. If Paxton drew upon his own work on greenhouses, his

Crystal Palace exterior. This contemporary view from the south side shows the transept or central hall with its high curving roof of iron arches and glass panes. From the transept extend the flat-roofed exhibition galleries of iron and glass. Lithograph by Day & Son, 1851. [Reprinted with permission of the Library of Congress P&P Division - cph 3g02906.]

methods were in use throughout industrial England, particularly in the great glass-and-iron train sheds of the London railway termini. The result was a marvelous freestanding structure of transparent walls and roof of glass, crystal to the Victorians, that was the size of a palace.

The scale of the Crystal Palace can be appreciated by looking at a controversy about its construction. As the exhibition hall was being approved for construction in Hyde Park, an argument arose that registers the nineteenth-century debate about the conflict of the machine and the natural world, resembling controversies about building the railway through the rural landscape. On the selected site were several venerable and stately elm trees. The question arose of whether the trees should be cut down to make way for a glass and iron hall designed to show off the machinery of the new era. As the sides quarreled, the matter was resolved by the decision to build the hall high enough to include the trees within the building itself. And so it came to pass. Three large elms were left to grow inside the building.

The ability to raise the roof to include existing trees suggests the vast scale of the Crystal Palace. The building was 1,848 feet long, the length of six American football fields, and 404 feet wide. In shape, the building took the form of a cathedral for the machine. There was a nave or flat-roofed corridor running the length, three times the length of St. Paul's Cathedral in London. The nave contained several floors. From the central corridor of the nave ran separate bays or avenues for display. This corridor was sixty-six feet high. At right angles to the nave at its center was a transept or corridor covered by a curved glass roof space 108 feet high that could accommodate the elm trees.

The glass and cast iron were manufactured off-site as interchangeable parts of standard sizes. As in the building of railway bridges and factories, cast iron was the crucial material for this large-scale construction. In the Crystal Palace, the several floors of the nave were supported by iron girders that rested on iron columns. In a nice touch of Victorian practicality, the iron columns were hollow to carry the rainwater run-off. The transept with its curved roof was supported, like the Iron Bridge at Coalbrookdale, by pre-fabricated iron ribs. With new efficiency, as many as 316 girders could be constructed in one week. At completion, the building used 2,300 girders and 3,330 cast-iron columns.

The use of glass panes to provide a transparent covering for the space was also made possible by new industrial methods. In the pre-industrial world, the material for glass had been made by hand, then in its molten form blown into shape by artisans working on a single piece. It was not feasible to make large sheets of glass in great quantities by these methods. Indeed, transparent glass for windows to let in the light had been a luxury item. In the 1830s, a method was introduced to make large transparent glass sheets for smooth panes quickly and in large quantities. The process still involved hand work, although routinized. A large ball of blown glass was rolled out on a metal table into the form of a cylinder. The sides of the cylinders were cut along its length to form a remaining sheet of glass which was then polished to a true face. For the Crystal Palace, this method was employed to produce the 293,655 individual panes of transparent sheet glass, each four feet by ten inches, a markedly large area for the time. In the completed building, these panes covered nine hundred thousand square feet, providing the entry of natural light to the exhibition space.

As a sign of the quickening pace of the machine age, the Crystal Palace, a building of the magnitude of the ancient cathedrals that had taken masons and carvers centuries to construct, was completed in six months. Possession of the site in Hyde Park was taken at the end of July 1850; the building was completed in January 1851. The reception of machinery and manufactures began in February. The exhibits were quickly installed and the interior prepared so that Queen Victoria could open the Great Exhibition to the public in May of 1851.

This quantum leap in speed of construction was due to the efficiency of industrial methods in producing and assembling reliable material in standard sizes. The work of building became, then, a work of putting together material made off-site. The method can be seen still in the construction of our own high-rise buildings that use prefabricated steel frames on which are hung prefabricated panels of glass or other materials. Prefabricated iron girders cast off-site were delivered to Hyde Park and tested for dimensions and strength. They were then raised by a steam crane to their positions, often in a matter of minutes. Similarly, the great curved iron ribs that spanned the transept roof were cut with steam-powered machines into equal sizes before being erected. With the metal forms of ribs, girders, and columns completed, workmen installed the glass panes using pre-cut metal sash-bars to hold the glass in place. A man could install as many as 108 panes in one day. The metal gutters that ran along the glass for drainage of rainwater were manufactured in long lengths, then cut to size on site. For the workmen at the Crystal Palace, then, the work of building consisted simply of piecing together the columns, girders, and ribs of iron, then fitting the innumerable panes of sheet glass. The labor resembled that of the assembly line in a factory.

The Crystal Palace manifested the emergent architectural style whose precondition is the rise of the machine. The style is generally termed functionalism; it has become the dominant mode of our own time. With a preference for new technologies, functionalism holds that a building should be designed primarily to carry out the functions or needs of that building. The style breaks with traditional architectural practice in its absence of decorative ornamentation. The Gothic cathedral may be highly functional, with stone arches that span a vast space for worship and flying buttresses on the exterior to hold the stress on the external walls. Yet it is ornamented.

Following tradition, many Victorians thought of buildings for public use as frames to be covered with symbolic ornament. Thus, attention was paid by the Victorians not only to structural elements, but also to the stone facing often covered with carvings of biblical scenes. Similarly, nineteenth-century public buildings in America had facades with stone carvings of symbolic figures such as Liberty or Justice.

In their construction, the structures of the machine age, such as railway bridges, train sheds, and factories manifest the emergence of the functional ideal, particularly in the adaptation of new materials and industrial methods. The strength of cast iron and the method of prefabrication were employed to create such railway structures as the great iron tubes that carried trains over the Menai Strait. Similarly, since cotton mills needed large open spaces to accommodate textile machinery as well as provide for the continuity of production from one stage to another, the owners turned to cast-iron beams and columns. Furthermore, since the factory was a for-profit operation, money was not spent on the merely decorative. The exterior and interior were made of plain brick. For the Victorians, these exemplary creations of the machine age were not considered architecture, but merely constructions that served certain purposes such as spanning rivers, sheltering railway passengers, or enclosing textile production.

The significance of the Crystal Palace is that these same functionalist principles were applied to a free-standing building intended to symbolize England's very identity and erected at the very center of the nation's capital. The instinct for practicality on the part of the Commissioners along with the need for an easily built, easily deconstructed, naturally illuminated, vast exhibition hall led them to reject the ornamented stone building of the past. They chose instead to erect a new kind of building using industrial products of cast iron and sheet glass manufactured in interchangeable modules to be assembled on the spot.

As in the case of train sheds, there was still some ambivalence about ornament. Rather than keeping the natural gray color of the material, the iron girders and columns were painted with red, light blue, yellow, and white. And yet, although the façade held no symbolic statues, the very absence of such traditional signs made the entire structure in itself a symbolic statement. The naked machine-made frame was a monument to the power and, in a way only partly realized by the Victorians, to the new beauty of a machine aesthetic. The Crystal Palace is now seen as a highly significant precursor of modern machine-based architecture.

NATIONAL FESTIVAL OF THE MACHINE

Queen Victoria and her Consort, Prince Albert, who was the driving force behind the Great Exhibition, opened the Crystal Palace on May 1, 1851, in a

rich and gorgeous ceremony whose sights and emotion are best described by
the queen herself:

> A little rain fell, just as we started, but before we reached the Crystal
> Palace, the sun shone upon the gigantic edifice, upon which the flags
> of every nation were flying.... The glimpse, through the iron gates of
> the Transept, the waving palms and flowers, the myriads of people
> filling the galleries and seats around ... gave a sensation I shall never
> forget, and I felt very much moved ... The sight as we came to the
> centre ... facing the beautiful crystal fountain was magic and
> impressive.[4]

Queen Victoria's intense admiration and excitement was shared by the
huge crowds at the opening. She notes that "the Nave was full of people ...
and deafening cheers and waving of handkerchiefs continued the whole time
of our long walk from one end of the building to the other." There was, she
writes, "joy expressed in every face."[5] The joy and the sense of magic continued in the crowds that flocked to the Exhibition. Like Victoria, the people
of England saw the Great Exhibition as an event that celebrated their nation
as the world leader in the industrial activity that defined the future, and that
had brought a profusion of material delights in the present.

The number of people who came to see the machinery and the manufactures within the great glass building was extraordinary. There was a virtual
frenzy in the nation to visit the Crystal Palace. The official tally showed
6,039,195 visitors through the mere five months the Exhibition was open.
The demand for admission was so great across the nation that the railways
commissioned special trains to bring people from throughout the country,
especially from the manufacturing cities of the Midlands, to London on special one-day excursions.

Nothing is more illuminating of the widespread awe of the machine and
the belief in the promise of machine-made prosperity than the social range
of the visitors. People of all classes came to the spectacle. One day a week
was set aside to accommodate factory workers who desired to see the
machines that provided their employment. On those special days the cost of
admission was dropped to the affordable price of one shilling. One contemporary middle-class commentator wrote with some surprise of the "industrial
classes" at the Crystal Palace: "[One] will see them well-dressed, orderly
and sedate, earnestly engaged in examining all that interests them ... playing
with manifest propriety and good temper the important part assigned to
them at this gathering of the nations."[6] That the appeal of the Great Exhibition cut across class lines suggests how the shift of wealth from agriculture
to industry and the increasing availability of machine-made possessions
tended to loosen the rigid class boundaries of Victorian England.

Machinery as Spectacle

The vast array of exhibits at the Crystal Palace were divided into four main sections: "Raw Materials," "Machinery," "Manufactures," and "Fine Arts."

The varied ways in which the machine had transformed Victorian life in such areas as transportation, building, warfare, and agriculture appear in the structuring of the displays within the "Machinery" section. The section called "Machines for direct use, including Carriages and Railway and Naval Mechanism" included the sleek, what we would call streamlined, steam locomotive called the Folkstone, resting on iron rails. "Civil Engineering, Architectural, and Building Contrivances" showed steam-powered cranes. "Naval Architecture, Military Engineering, Ordinance, Armour and Accoutrements" displayed machine-made instruments of war employed to expand the British Empire. Here was a model of the warship HMS *Medea* with a combination of sail with a steam-driven side paddle-wheel. With pride, the official catalogue stated, "This splendid war steamer has the reputation of being one of the fastest paddle-steamers under canvas in the Royal Navy." This ship was renowned for bringing the Koh-i-noor diamond, one of the largest diamonds ever found, from India where it had been seized by the British to England where it became part of the Crown Jewels.

"Agricultural and Horticultural Machines and Implements" demonstrated the emergent industrialization of agriculture. Iron devices such as plows, wagons, and steam-powered threshers were displayed in a large factory-like space within the Crystal Palace. On the grounds outside the building, agricultural machinery was shown at work. Here, the chief attraction was a steam-powered version of what we would now call a tractor. This small-scale steam locomotive with boiler and smokestack ran over the ground on four large wheels without the need of tracks. This Victorian tractor, what the catalogue calls a "small portable steam engine,"[7] was a radical innovation in replacing the age-old use of animals to pull agricultural implements such as plows or reapers. Here, steam locomotion was expanded to pulling iron machines for planting and harvesting over uneven, muddy ground. Furthermore, the steam tractor was prophetic of the use of machine power for self-propelled motion without tracks. Some of the first automobiles were powered by steam.

The implements pulled by the steam tractor also brought industrial innovation to the ancient practices of farming. Now mass-produced from cast iron, the new plows, planters, and threshers were inexpensive and long-lasting. As in the mechanization of textile production, the development of large-scale machines brought increased efficiency by bringing together in one mechanical device separate activities formerly done by hand. The "Machinery" section showed the new "combine" developed by Cyrus McCormick in

America that combined the separate activities of reaping, threshing, and binding grain into one continuous process. As the catalogue notes of the McCormick combine, "In agriculture, it appears that the machine will be as important as the spinning-jenny and power-loom in manufactures."[8] The Victorians realized that the agricultural machinery shown at the Crystal Palace prophesied a new era in farming methods and in rural life: "The most remarkable feature in the agricultural operations of the present day is undoubtedly the rapid introduction and use of small portable steam engines for agricultural purposes, especially noticeable in connection with the combined threshing, straw shaking and dressing machines, unknown until within the last two years."[9]

For the thousands of visitors, the most attractive part of the "Machinery" section, perhaps of the entire Crystal Palace, was the area of "Manufacturing Machines and Tools, or Systems of Machinery, Tools, and Implements" more commonly known in its time as the Hall of Machinery. Here, machinery at work became spectacle. Brought out of their hidden life in factory and workshop, the "Manufacturing Machines" themselves that were the source of England's new prosperity were put on view for the British public. The Hall of Machinery was seen as the very center of England's essential identity as the workshop of the world. As a contemporary wrote,

Great Exhibition, Hall of Machinery. In this popular gallery of iron and glass were exhibited such machines as a hydraulic press; a locomotive; James Nasmyth's steam hammer; steam-powered cranes; a turntable; and a steam pump. From Dickinson, Comprehensive Pictures of the Great Exhibition of 1851 (1854). [Reprinted with permission of The British Library Board - Shelfmark Cup.652.c.33 volume 2, plate XXII.]

"In this Department, which might well be called the great park of machinery, a thoughtful observer could easily discover the distinctive character of the English nation, as regards political economy—Englishmen employ their capital, but are ever seeking for mechanical means to work it."[10]

Here were machines in operation. There were "thousands of little machines, which well deserved the epithet of beautiful ... hard at work and ingeniously occupied in the manufacture of all sorts of useful articles from knife handles to envelopes."[11] Hard at work, weaving densely patterned wool damask, was a sizeable jacquard loom of the type that inspired Charles Babbage's proto-computer, as well as small mechanical looms. There was an envelope-folding machine at work showing how a single operative could produce twenty-seven hundred envelopes per hour. There was a cigarette-making device that made eighty to one hundred cigarettes per minute. There was a model of a patent steam traveling crane. There were the grand instruments of construction such as the steam hammer and the forging iron. Quite prominent was the massive hydraulic press that had forced into shape the links of the great iron chains that held in place the iron tubes of the Britannia Bridge carrying the railway over the Menai Strait.

For the multitude who thronged the Hall of Machinery, the sight of the massive mechanical devices set in place and fully operating was truly a source of wonder. Eager to show machinery in action, the builders had included within the Crystal Palace complex a small external engine house containing stationary steam engines similar to the engine house outside a factory. These engines generated steam that ran in pipes into the Crystal Palace itself and then below the floor of the building to the machinery. Such care indicates the Commissioners' sense of the centrality of the machine itself to the Exhibition as well as the desire to present such machinery in its true state of motion. The organizers were, as the *Official Catalogue* of the exhibition notes, intent on showing the "overwhelming impression of speed and power in industrial machinery.[12]

In the eighteenth century, commercial exhibition halls had offered paying customers the opportunity to see the wonders of clockwork machinery in action, as in Merlin's Mechanical Museum that featured clockwork human figures or automata.[13] With its mechanical imitations of human actions, Merlin's was a favorite haunt of Charles Babbage. The Hall of Machinery in the Crystal Palace expanded this sense of machinery as spectacle in a mode suited to the heroic scale and grandeur of steam-powered devices. Henry Mayhew, a noted journalist of the age, wrote a mild social satire in *The World's Show, 1851, or, The Adventures of Mr and Mrs. Sandboys and Family, Who Came Up to London to "Enjoy Themselves" and to See the Great Exhibition*. He provides a most telling account of the delight of the public in the spectacle of the machine in operation. Here we see the Victorian passion for the new machinery that cuts across class lines to include those who entered at the day of a reduced rate, such as the "railway guard," the "carpenter," the "Blue-coat boy" [charity school student], and the "Life-Guardsman" [soldier]:

> But if the other parts of the Great Exhibition are curious and instructive, the machinery, which has been from the first the grand focus of attraction, is, on the "shilling days," the most peculiar sight of the whole.... The people press, two or three deep, with their heads stretched out, watching intently the operations of the moving mechanism. You see the farmers, their dusty hats telling of the distance they have come, with their mouths wide agape, leaning over the bars to see the self-acting mills at work, and smiling as they behold the frame spontaneously draw itself out, and then spontaneously run back again....
>
> The chief centres of curiosity are the power-looms, and in front of these are gathered small groups of artisans, and labourers, and young men whose red coarse hands tell you they do something for their living, all eagerly listening to the attendant, as he explains the operations, after stopping the loom.... Indeed, whether it be the noisy flax-crushing machine, or the splashing centrifugal pump, or the clatter of the Jacquard lace machine, or the bewildering whirling of the cylindrical steam-press—round each and all these are anxious, intelligent, and simple-minded artisans, and farmers, and servants, and youths, and children clustered, endeavouring to solve the mystery of its complex operations.[14]

As a middle-class observer who wrote extensively of the London poor, Mayhew is gratified at the pleasure in the machine expressed by the working classes of England.

If the machines were indeed a show in themselves, the organizers of the Exhibition recognized that the transformative energy of the machine age came not only from individual mechanical devices, but, as the title of the exhibition section clearly indicates, "Systems of Machinery." The grand design of the Great Exhibition, with its inclusion of an engine house to provide built-in steam power, testifies to the desire to display machines not as static objects, but in motion within a system of machinery. Quite brilliantly, the Commissioners transported and set to work in the center of London the exemplary system of machinery of the time in the coordinated machines of the cotton mill. In a single vast space, fifteen transported machines were powered as in a textile mill by belts attached to overhead spinning rods turned by the steam produced in the external engine house. The ensemble of machines opened up the raw cotton, then carded, spun, and wove the raw material to produce the finished textile in a coordinated linear process. In a scene reminiscent of our own science museums where school tours watch contemporary technologies at work, middle-class Victorian families, the women in crinolines, the men in frock coats and top hats, stood for their safety behind a cast-iron railing (no danger of being ground in the mill for the visitors to the Crystal Palace), watching intently as the system of machines manufactured finished textiles.

We might inquire, then, as to sources of the near-universal sense of wonder among all classes at the "self-acting mills" and the "power-looms," the "cylindrical steam-press" and the "Jacquard lace machine" and "flax-crushing machine," and the mechanics of producing cotton fabric. One reason, of course, was the sheer newness. Except for the industrial workers in the Midlands, most Victorians such as middle-class tradesman or rural aristocrats had never seen the steam-powered mechanical devices that were transforming their world. In the early Victorian period there flourished what can be called techno-tourism, in which people traveled to the industrial cities in the Midlands and even to the naval armory at Portsmouth to satisfy their curiosity about the new machines. With the Great Exhibition, such travel was no longer necessary, at least for Londoners; the factory and the mill had been transported to Hyde Park. A modern analogy might be the crowds that gathered to watch the launch of rockets from Cape Canaveral in the heady early days of the United States space program. Or the crowds that make the National Aeronautics and Space Museum, with the first space capsule, the most visited museum in America.

And there was also the Victorian interest, now somewhat lost, in just how things work.[15] Visitors delighted in learning how things were made by seeing the process by which machinery transformed raw material into finished goods. Perhaps this enthusiasm was due to a transparency in the operation of the Victorian machine in contrast to contemporary technology. There is little of interest in watching a computer work. In a computer, all the activity takes place within the black box interior where the movement of electrons in silicon chips is invisible to the eye. But the Victorian machine showed the clearly visible motion of large parts moving in rhythm. Rotating rods turned leather belts. Moving cylinders carded the raw cotton. Shuttles flew through the loom. Gears meshed and finished, patterned cloth emerged. There was a powerful visual fascination to Victorian machinery such as the working loom or the rushing locomotive.

COMMODITY CULTURE AND THE MACHINE

The display of "Machinery" with its many sub-sections demonstrated the process of making by machine. It showed to the general Victorian public the usually hidden work of systems of steam-powered machinery. These displays demonstrated to admiring crowds the power, the productivity, the cleverness, and the particular beauty of these machines in motion. Such machinery as the hydraulic press, the steam tractor, and the locomotive were valued for their usefulness. They forged massive chain links out of cast iron, harvested wheat, and pulled carriages for the public conveyance. For such machines, worth was not defined by their monetary value but by their efficacy in performing their set functions. Their purchase price was determined by their utility. The Victorian spectators in the Hall of Machinery were occupied not with the monetary value of the machines, but with their efficiency in performing their tasks.

In contrast, the gallery of "Manufactures" at the Great Exhibition displayed the products made by the machine within the factory system. A term coined in the machine age, manufactures refers to products brought into being from raw material by large-scale machine production. Like the related term manufacturing, the word manufactures became necessary in the transition from handwork to machine work. Although the Crystal Palace did display hand-made goods, chiefly from non-industrialized nations within the Empire such as India and Australia, the primacy in the Exhibition given to "Manufactures" indicates the emphasis on celebrating the productivity of the machine.

The "Manufactures" in the Crystal Palace were primarily machine-made goods such as chairs, knives, and metal dinner plates intended for purchase by individuals for their private use, primarily in the home. These saleable items have come to be called commodities, in the sense of objects whose primary worth is defined by their monetary value. These goods made by the machine may have a utilitarian function. A chair does provide a place to sit. But the value of the Victorian mass-produced chair with its arms and back carved by machinery and covered in machine-woven cotton is not defined primarily by its function. The value of the machine-made chair depends in part on the cost of its making, but primarily on its decorative qualities, particularly its fashionability. To use nineteenth-century terms, the value lies not in use, but in cash exchange. Thus, the "Manufactures" section of the Crystal Palace displays with pride not the machines themselves, but their products as available for consumption.

The separation in the Crystal Palace of the Hall of Machinery from the "Gallery of Manufactures" gave physical form to an issue in thinking about the machine age that continues into our time. One result of this physical division was that looking at the commodity such as the chair or the knife removed from the process by which it was made tended to keep from consciousness the work and workers in the factories that produced these objects. In our world, the weakening of that interest in making that the spectators at the Crystal Palace demonstrated has been replaced by virtually complete focus on manufactures in themselves. The Great Exhibition, then, can be seen as the first celebration of what has come to be called the culture of the commodity.

The emergence of the culture of the commodity is exemplified in the "Carriage Section," which was the Victorian equivalent of the modern auto show. The "Carriage Section" consisted of a large open space jammed with the four-wheeled vehicles so sought after by the Victorian middle class. Since the components of the vehicles such as metal-spoked wheels and iron springs could be mass-produced, carriages had become widely affordable and London was troubled by that most modern phenomenon, the traffic jam. In part an object of utility, the carriage had also become a commodity defined by its monetary value. The type of carriage a Victorian drove was as much a sign of social status as a specific kind of automobile in contemporary America.

The "Gallery of Manufactures" and the "Carriage Section" of the Crystal Palace provided a setting for the act of purchasing as the new personal drama of all social classes in the machine age. Identity became defined by the commodities we buy and own. The productivity of the machine had generated a society of consumers. The nature of the new consumer society generated by the rise of machine can be neatly traced in the history of the word "consume."[16] In pre-industrial society, the term consume implied a form of destruction, as in consumed by fire, or in the labeling of tuberculosis as consumption. By the eighteenth century, the term came to be applied to the using of goods and services. In the nineteenth century, the word consumer gradually replaced "customer" as the nature of buying was perceived as moving from personal interaction with an individual seller to a more impersonal practice of purchasing mass-produced commodities. In our own society, we have come to speak of the populace less as citizens defined by their participation in society than as consumers whose identity lies in their monetary choices about buying that determine the state of the economy and of the nation.

The transformation of citizen to consumer that accompanied the rise of the machine appeared at the Great Exhibition in the delight of visitors of all classes in the profusion of machine-made commodities ranged before their eyes. In pre-industrial agricultural society, wealth and status were determined by the possession of arable land for farming. In England, such land had remained into the nineteenth century for the most part the possession of a hereditary aristocracy. With the rise of industrialism, enormous wealth had been acquired by another small group, the owners of the factories and textile mills. The display of manufactures at the Great Exhibition demonstrated that a new form of wealth, as measured by the possession of material goods, had begun to spread widely through the society. The machine had made commodities inexpensive and available to all with the ability to pay. Elaborate furniture of carved wood, silver candelabra and serving dishes, upholstered furniture, and brightly colored drinking glasses had once been made by hand and thus been extremely expensive, with their possession limited to the aristocracy. Now that these objects were mass-produced they could be purchased by the middle class and even by the working class. Such wares could be bought with the wages generated by the same industrial revolution that had made such commodities affordable. The ostentatious display of such material possessions in the overcrowded Victorian parlor became the source of social status for the middle and working classes.

As much as the Great Exhibition displayed thousands of machine-made goods for personal consumption, it was strictly forbidden to sell the manufactures within the Crystal Palace itself. The official guide states, "No article is allowed to be sold in the Building, except the Official Catalogues, the Medals struck at the press, refreshments, and bouquets of flowers." Unlike our own trade fairs, companies were not permitted to take orders for their plated silverware, jacquard damask tablecloths, or their machine-carved wooden chairs. This restriction on commerce within the Crystal Palace itself represents the Victorian ambivalence toward consumption. Delight in acquiring machine-made goods mixed with the Victorian ethical doubts about making such acquisition central to personal life. The Exhibition organizers wanted not only to display the power of British industrialism, but also to ennoble industry as the working out of divine providence and as a testament to the work ethic of the British laborer. To do so, they had to banish from the glass building the main motivation of manufacture, commercial gain.

Given the plenitude of machine-made commodities, the organizers had to develop a set of categories for the exhibition of manufactures to provide a spatial order for the psychologically overwhelming display of machine-made goods. The exhibition constructed two broad groupings: "Textile Fabrics" and "Metallic, Vitreous and Ceramic," each divided into subdivisions. The multiple display spaces of the "Textile Fabrics" class showed pride in the application of the system of machines as displayed in the Hall of Machinery to a range of raw material. There were "Cotton Manufactures"; "Woolen and Worsted Manufactures"; "Silk and Velvet Manufactures"; "Manufactures from Flax and Hemp"; "Mixed Fabrics, including Shawls, but exclusive of Worsted Goods"; and "Articles of Clothing for immediate Personal or Domestic Use." The division "Metallic, Vitreous and Ceramic," included a broad range of goods primarily for domestic and personal use manufactured from various materials. The "Metallic" objects included such categories as "Cutlery and Edge-Tools" and "Iron and General Hardware." The "Vitreous" or glass subdivision, echoing the glass panes of the Crystal Palace itself, incorporated "Stained and Painted Glass, Optical Glass, etc." The "Ceramic Manufactures" showed "China, Porcelain, Earthenware, etc."

In the abundance of objects blazing forth the prowess of mechanized production appear two major themes still at work in our own society. First, in an analogy with the replacement of animal and muscle power by steam power, the manufactures displayed the replacement of handwork by the skill of the new machines working without human guidance. A prime example was the carving of wood into elaborate patterns by machine. Second, the manufactures showed off the replacement of natural materials by innovative substances that resemble the natural, much like contemporary plastic can be made to look like wood grain. For the crowds at the Crystal Palace, the substitution of the machine for the hand- and man-made material for the natural was the source of widespread if not universal admiration.

In showing off the virtually miraculous results of industrialization, the Great Exhibition tended to highlight not the ordinary, but the extraordinary. There was, for example, deep interest in the application of machine techniques to areas of textile production traditionally associated with the most artistic and the most demanding handwork. For example, traditionally, damask, a cloth of cotton or silk with an elaborate reversible pattern often used for table linen, had been fabricated by skilled hand weaving and thus was an expensive possession available only to an aristocratic elite. The exhibition featured damask made by the jacquard loom that could weave elaborate patterns programmed by punch cards. The colored tablecloths and table napkins manufactured by a Scottish firm, for example, were praised both in themselves and implicitly as signs of the skill of the machine at weaving these elaborate patterns: "The border is Gothic, having the figures of St. George, St. Andrew, and St. Patrick in the corner niches, with St. George and the Dragon in the centre."[17] The *Art Journal*, the leading art magazine of the time, particularly praised the way that the jacquard loom had now brought such beautifully woven material within the purchasing power of all classes. The jacquard could be "used for every description of cloth, from that made expressly for her Majesty's table, to the coarsest 'whitey-brown' destined to cover the pine-board of some American backwoodsman."[18]

The "Textile Fabrics" section also proudly offered examples of machine-made lace. Lace is a complex fabric with open spaces of netting holding forms of solid needlework. Because of the demanding and tedious hand labor involved, lace had been a luxury item, highly expensive and much in demand. In response to this demand, steam-driven lace-making machines were developed in the nineteenth century that were themselves highly complex. The machine-made lace became very popular and in the Crystal Palace an exemplar of the progress of industry. The Great Exhibition catalogue noted that in the lace on display "both the ground and the pattern have been entirely produced by the machine."[19] A white lace scarf from "an extensive manufacturer of lace" was unabashedly praised by the *Art Journal* as an "imitation of Brussels point,"[20] a complex traditional method of lace-making. Again, a luxury item formerly available only to the wealthy and to such wealthy institutions as the Anglican and Roman Catholic Church had been transformed by the machine into a commodity obtainable by the middle and working class.

Within the "Metallic, Vitreous and Ceramic" section, the subdivision of "Cutlery and Edge-Tools" showed off the metal knives and tableware of Sheffield, a Midlands city specializing in such products. The efficiency of concentrating similar factories, suppliers, and skilled labor had led to such specialization in Victorian manufacturing towns. Queen Victoria was especially impressed by her "very detailed inspection of the Sheffield ware, beginning with a model of the process by which steel is made from iron and then finished into knives.... There were 'Bowie' knives in profusion, made entirely for Americans, who never move without one."[21] This cutlery section showcased the multitude of diverse utensils beloved by middle-class Victorian diners—fish knives and fish forks, fish carvers, game carvers, dessert knives and dessert forks, as well as straight razors for the Victorian gentleman and scissors for the Victorian lady.

Like the train sheds and the Crystal Palace itself, the cutlery of Sheffield mixed the highly functional materials of the machine age with an elaborate decorative scheme. The objects were shaped both by traditional handwork as well as machine manufacturing. Blades were no longer hammered on the anvil by the blacksmith, but honed by a steam-powered blade-making machine. But the table knife would have an ivory handle with inlays done by hand. In the straight razor, the ivory handles as well as the steel blades were ornately engraved. Often, ornament symbolized function. Thus a game-carving knife had a hand-engraved design of dead game birds in the ivory-and-silver handle that held a "blade of the highest polished steel."[22] The blade of a fish knife was engraved with a heron wading in a stream and looking for his own meal of fish. Some examples showed function run amok. One of the standouts in the Crystal Palace was a sportsman's folding knife by J. Rogers & Sons, with eighty blades and instruments such as a corkscrew, awl, saw, pincers, and hooks—the Swiss army knife of its day. Such a non-functional epitome of function, with each blade and each tool individually engraved, appears now as a bravura show of the virtuosity of machine manufacture.

The manufactures at the Crystal Palace celebrated not only the elaborateness and cheapness of commodities made by machine, but also the transformation of the very material from which these goods are made. Industrial innovation freed manufacturing from dependence on natural materials such as wood, or refined metals such as silver, or expensive animal parts such as ivory. As we now use nylon or plastic, Victorian manufacturing moved to the use of synthetic materials on an industrial scale, thus lowering the cost of commodities and expanding the boundaries of consumption. As the

Victorians delighted in how machines worked, so they enjoyed the cleverness in the substitution of synthetic for natural materials.

The fish knives and game carvers, as well as the teapots, cruets, vases, and candelabra at the Crystal Palace only appeared to be silver. Rather than pure silver, they were imitations with a silver exterior and an iron center. Most of the metal objects in the Great Exhibition were manufactured by the new and increasingly popular method of electroplating. Electroplating is a process by which a thin coating of metal is applied to the surface of another metal by the action of electricity. The object to be coated is put into a solution that contains the dissolved metal to be applied. The object then is given a negative electrical charge and the dissolved metal, which has a positive electrical charge is attracted to the surface, where it adheres in a thin film or plating. Developed in Europe in the early nineteenth century, the process used a voltaic pile to run the electrical charges through the objects. By 1839, scientists in Britain devised processes for the copper electroplating of printing press plates. By 1840, John Wright, of Birmingham, had developed a method for gold- and silver-electroplating. His associates, George and Henry Elkington, were awarded the first patents for electroplating in 1840. As industrial cities specialized, Birmingham became the center of the electroplating industry. Commercial electroplating of nickel, brass, tin, and zinc was developed by the 1850s. The process was applied on a large scale with the development of electrical generators and the distribution of electricity from central generating stations in the latter nineteenth century.

At the Crystal Palace, the Elkington firm presented a variety of objects in electroplated silver. Exemplifying the Victorian delight in innovation, the *Art Journal* praised the firm's "celebrated electro-plate manufactures, a branch of industrial art which has made immense strides since the patent for the various processes of gilding and plating materials by the agency of electricity was granted to this firm in 1840."[23] To the age, one of the "advantages which electro-plating possesses" was the somewhat paradoxical ability to employ industrial methods to reproduce efficiently and cheaply commodities that imitated hand-fashioned styles of the past. Exemplifying the "variety and artistic qualities of their [Elkington's] productions" were fruit dishes "ornamented with subjects taken from the English nation's games—cricket and archery"; a communion service in the "ornamented Gothic style"; a tea service in an elegant adaptation of the arabesque pattern"; and a many-armed candle holder in an organic form "modelled from the 'Crown imperial' plant."[24]

The widespread use of another substance, papier-mâché, at the Great Exhibition provides an example of the Victorians finding innovative uses for older materials. In papier-mâché the Victorians did not invent a new substance, but rather adopted an older craft tradition for industrial production to create an inexpensive and malleable substance. The term, literally the

French for "chewed paper," because of its appearance, dates back to ancient China. By the end of the eighteenth century, a process for using papier-mâché on a large scale was developed and patented in Birmingham, which remained the center of such manufacture. The making of papier-mâché, however, indicates the uneven developments in manufacturing in the nineteenth century, whereby machine production coexisted in the factory setting along with routinized, often backbreaking physical labor done by poorly paid workers. In industrial manufacturing, papier-mâché begins as sheets composed usually of fabric scraps held together by glue. The invention of a continuous sheet paper machine allowed paper sheets to be made of any length and the sheets bound to any thickness determined by the number of sheets. One of the first composite materials, papier-mâché resembles our own plywood made of bonded layers of thin-cut wood. Although the sheets of papier-mâché could be produced quickly by machine, workers had to physically dip the large sheets in shallow vats of tar spirit and linseed oil, then paste them over greased molds. The papier-mâché object would be built up in this way, continually adding layers and baking until the final thickness, sometimes as much as one hundred layers, was met. Since the material was wet and malleable when formed, it could be applied to a mold so as to take the form of the mold when dry. With its strength and malleability, papier-mâché was the plastic of its day.

Papier-mâché had been first used in the early Victorian period for small objects such as dolls, trays, workboxes, inkstands, face screens, snuff boxes, and letter holders. By the time of the Great Exhibition the industrial innovations in its manufacturing had made possible the construction of larger domestic items such as dressing tables, wardrobes, and beds. To make a chair, for example, layers of papier-mâché were pasted onto a wooden or metal base, each layer being polished and smoothed to give an even surface, then decorated. The Crystal Palace showed an easy chair called the Day Dreamer with buttoned upholstery on a papier-mâché frame. The frame with curved arms and legs appeared to be elaborately carved wood, but was actually molded from malleable papier-mâché. In keeping with the name of the object, the frame was decorated with figures of nude women that supposedly fill the daydreams of the seated male.

This malleable material could be used to produce a multitude of commodities. The display of one prominent Birmingham manufacturer included an inkstand in black with a stag flanked by two hunting dogs as well as a baby bassinet in the shape of a nautilus sea shell ornamented with flowers. Contemporary comment again indicates the Victorian praise of innovation for increasing the flow of domestic commodities: "The manufacture of papier-mâché into a great variety of useful articles of large size is the result of the efforts made, within a comparatively recent period, by the various artisans who have devoted their attention to this important branch of the industrial arts. It is not many years since the limits of the trade were circumscribed to a tea-tray, but now we find articles of furniture ... of a more substantial kind, in chairs and sofas for the drawing room, or the entire casings of pianofortes."[25]

The Experience of the Commodity

The nationwide desire to visit the Crystal Palace was extraordinary. The range, variety, and number of the machines and commodities displayed were immense. But the crowds, machines, and manufactures have come down to us only as engravings in catalogs. What of the felt experience of the multitude that surged through the halls? What of the inner feelings of the people of all classes who witnessed the displays? More generally, what was the response of the first people who experienced fully the machine-made commodity culture?

From the contemporary accounts of visitors, the sheer newness and abundance could not be comprehended fully. As in Dorothy Wordsworth's first contact with the steam-powered pumping engine, visitors could not fully understand their contact with the machine age. In particular, they could not encompass the profusion of the machine-made products on display. The Victorians experienced the stimulation of machines and commodities within a sense of the dreamlike. Like most visitors, Charles Dickens was simply overwhelmed, "I find I am 'used up' by the Exhibition. I don't say 'there is nothing in it'—there's too much. I have only been twice; so many things bewildered me. I have a natural horror of sights, and the fusion of so many sights in one has not decreased it."[26] William Makepeace Thackeray, the prominent Victorian novelist, wrote a sly poem, "Mr. Maloney's Account of the Crystal Palace," in the dialect of a working-class visitor who also could not fully comprehend the flood of commodities and "was dazzled quite/ And couldn't see for staring."[27]

Perhaps in our own time of big-box stores, TV ads, and electronic billboards we have become accustomed to the plenitude of the commodity. Perhaps we have developed psychic defenses against the assault on the senses that comes with living within the commodity culture.

4

Shrinking the Globe, Expanding the Empire

With the rise of the machine, the pace of invention quickened through the nineteenth century. The factory system transformed the production of goods and the nature of work. A new machine-based architectural style of iron and glass radically altered the shapes of buildings and continues into our own time. The emergence of inexpensive manufactures generated the consumer culture that still dominates our lives. This chapter will consider the global effects of such technological innovation.

From the 1840s, England began to be knit together by the railway and the electric telegraph. With the progressive dynamic of technological change, the expansion of these inventions continued in a nineteenth-century form of globalization. The steam-powered iron locomotive moved onto a global scale as the ocean-going steam-powered iron steamship. The electric telegraph lines through which flashed orders for the railways of England became the undersea electric telegraph cable spreading over the world. What we might call space–time was transformed by these innovations. Distance considered as a function of time radically decreased with steamship travel between London and India and between Southampton, England and New York. With the opening of the transatlantic cable and of cable service between London and New Delhi, time considered as time experienced in the sending and receiving of messages seemed to have been obliterated as electrical impulses moved words that could be read almost instantaneously. Just as the steam engine, the power loom, and the railway operate within a system, so the undersea cable and the steam-powered ocean liner also function within a system, the system of global communication and of empire. It is only for the purposes of discussion that the components are considered separately.

THE TELEGRAPH

In many ways the electric telegraph was as crucial a device to the Victorian age as the steam engine. The steam engine developed through incremental technical improvements by skilled artisans and experimenters such as James Watt. In contrast, the telegraph grew from developments in what we would now call science and from the laboratory rather than the workshop.

The electric telegraph emerged from investigations early in the nineteenth century into the phenomenon of electromagnetism. Electromagnetism describes the observable fact that an electric current can induce a magnetic field in an iron object and that, conversely, a magnet can induce an electric current in a wire. A changing magnetic field produces a changing electric field; a changing electric field generates changes in a magnetic field. In 1820, it was discovered that an electric current produced by a chemical battery produces a magnetic field which will deflect a compass needle. Also in 1820, the galvanometer, a coil of wire around a compass, was developed which could be used as a sensitive indicator for electric current. In the later 1820s, it was found that winding an electrically charged wire around a core of iron produced an electromagnet that could function even through the high resistance of long telegraph wires. (A more detailed account of British contributions to the applications of electromagnetism appears in the "Generation of Electricity" section of chapter 6, Late Victorian Invention.)

Basically, then, the electric telegraph operates on a simple principle based on electromagnetism. Wires are stretched between points of communication and an electric current is sent through these wires by a battery. The telegraph key, at first consisting of two pieces of brass or copper, can be pressed together to complete the electrical circuit or allowed to spring apart to break the circuit. At the other end of the wire from the key is an electromagnet consisting of a coil of wire which becomes activated when a current is passed through it. In the early electric telegraph, running a current through the wires would activate specific electromagnets that would swing compass needles to point to letters on a board.

The first commercial electrical telegraph was patented by Sir William Fothergill Cooke in 1837, the first year of Victoria's reign. The telegraph was particularly important to the railways since the wires could carry messages at speeds that outraced the trains to transmit information of arrivals and departures as well as to synchronize railway time. Needing their own telegraph lines, the railways faced major construction projects. But the practical problem of stringing long lengths of wire to carry the current turned out to be relatively easy to solve. Since the new railways had acquired the right of way for their lines, the construction of the telegraph lines was accomplished by running the wires alongside the tracks. The railway telegraph first entered use on the Great Western Railway in 1839, running thirteen miles from Paddington Station in London to West Drayton. In 1842, the line was extended twenty miles west of London, and by mid-century, telegraph lines ran alongside all the railways of England.

In the early 1840s, an American, Samuel F. B. Morse, developed a system quickly adopted by the British that replaced the cumbersome needle telegraph and continued in use through the nineteenth century and into the twentieth. Morse used an electromagnet to move a pencil to mark a moving strip of paper with short and long marks depending on whether the key was held closed for a short or a long time. When the key was closed for a short time and then a longer time, the pencil marked the paper with a dot followed by a dash and this signified the letter "A." He assigned a combination of short and long symbols to each letter in the alphabet to form the Morse code in use until recent times.

In the 1850s, telegraph operators began to realize that they could recognize the different sounds made by dots and dashes. A new detector mechanism, called a sounder, was invented. This device used an electromagnet to pull on a piece of iron and make a clicking sound. When the electromagnet released the iron, it made a thinner- and lighter-sounding click. Operators learned to discriminate between these two sounds and to use this ability to tell whether they were hearing a dot or a dash. This method of interpreting the code by dots and dashes persisted well into the twentieth century.

Even beyond the railway, the electric telegraph was enormously successful in transforming many forms of English life. Like most inventions, the telegraph was an object of wonder and notoriety in its early days. In 1845, John Tawell was apprehended following the use of a needle telegraph message from the town Slough to Paddington Railway Station. This was hailed as the first use of the telegraph to catch a murderer. The message was: "A murder has just been committed at Salt Hill and the suspected murderer was seen to take a first class ticket to London by the train that left Slough at 7.42 PM. He is in the garb of a Kwaker with a brown great coat which reaches his feet. He is in the last compartment of the second first-class carriage."[1] The reason for the misspelling of 'Quaker' was that the British system did not support the letter Q.

To the Victorians, the phenomenon of what appeared to be instantaneous communication appeared to have eradicated space and time. As with the marvels of machinery at the Crystal Palace, the early telegraph became spectacle. A demonstration of the new telegraph in 1845 took the form of "an Exhibition admitted by its numerous visitors to be the most interesting and attractive of any in this great Metropolis.... Questions proposed by visitors will be asked by means of this apparatus and answers thereto will instantaneously be returned by a person 20 miles off, who will also, at their request, ring a bell, or fire a canon."[2] The capability to make virtually immediate contact with another person "20 miles off" and to transmit an order of action to "ring a bell" was not merely a trick, but marked the electric telegraph as a transformative innovation.

Furthermore, the telegraph made possible the national standardization of time to what the Victorians called railway time. With the speed of telegraphic communication each place along the ribbons of copper wire connecting stations could synchronize its time. This time was by agreement keyed to astronomical observations at the Royal Observatory in Greenwich, England, and called Greenwich Mean Time or GMT. This standard was adopted by all railway companies by 1848. Thus time was regularized for all, no longer ascertained by the clock on the village church but set to the standard GMT through a signal passed to the nation by the web of the electric telegraph.

The effect of the electric telegraph ran not only to the coordination of the railways, but also to the operation of the intensified commercial life of industrialization that increasingly depended on this mechanical innovation. With rapid communication of prices and of news affecting business, the British stock markets could buy and sell bonds and equities in real time. Thus, financial markets became truly national and even international, operating around the globe in real time, much as the international stock markets do today. In the interconnection that characterized the age of invention, telegraphic communication with its rapidity increased the scale of commerce and finance and thus enlarged the amount of capital available for further innovation.

As a device enabling a mysterious unseen energy to pulse through copper wires and allow an almost magical connection between minds over long distances, the electric telegraph provided a theme for the Victorian imagination. Alexander Somerville writes optimistically of a "moral electricity" uniting the aristocracy, workers, and capitalists, "moral electricity ... to carry the instantaneous message of one feeling, one interest, one object, one hope of success from the lordly end, to the working man's end of the social world."[3]

The electric telegraph was also quickly linked by the Victorians with erotic energy. There is an amusing account of "electric affinities" in a poem of the 1860s composed by none other than James Clerk Maxwell, the British scientist who contributed significantly to electromagnetic theory. "Valentine by a Telegraph Clerk to a Telegraph Clerk" incorporates technological details of the telegraph into a love poem as the male lover sends the message that the "tendrils" of his female lover's soul down the line "in close-coiled circuits wind/ Around the needle of my heart." He plaintively asks,

"What current is induced in thine?/ One click from thee will end my woes."[4]

Always receptive to technological change, Charles Dickens, the great Victorian novelist, incorporated the electric telegraph into his imagination. In his private memorandum book he connected the operation of the electric telegraph with the narrative method of his novels and his own work as a novelist: "Open the story by bringing two strongly contrasted places and strongly contrasted sets of people into the connexion necessary for the story, by means of an electric message. Describe the message—be the message—flashing along through space—over the earth, and under the sea."[5]

WIRING THE WORLD: THE VICTORIAN INTERNET

As we look back to the age of invention, our present technologies shape our sense of the technological past. The nineteenth century now appears to be significant in the history of technology not only as giving rise to the factory system and the consumer society, but also as bringing about the technologies of intelligence that dominate our own era. The nineteenth century brought not only the railway, but also the railway telegraph, not only the ocean-going steamship, but also the telegraph cable that moved words under the sea. The dominant communication technologies of our time were set in place during the Victorian period in the telegraph and the submarine cable: "There was ... an Internet.... During Queen Victoria's reign, a new communications technology was developed that allowed people to communicate almost instantly across great distances.... A worldwide communications network whose cables spanned continents and oceans."[6] Indeed, the purportedly stodgy Victorians were really wired. Late in the nineteenth century came the completion of the astonishing engineering feat of constructing the undersea cable system to wire together the world. Our e-mail now speeds through the material base of undersea wires laid down by the Victorians: "The world has actually been wired together by digital communications systems for a century and a half. Nothing that has happened during that time compares in its impact to the first exchange of messages between Queen Victoria and [American] President Buchanan in 1858."[7]

In 1858, through the completed undersea cable, Queen Victoria sent a telegram of congratulation to President Buchanan expressing the hope the cable would prove "an additional link between the nations whose friendship is founded on their common interest and reciprocal esteem." The president responded with the conventional yet sincere nineteenth-century rhetoric that envisioned the technological progress of England and America as the signs of divine favor, compared the efforts of technology as the moral equivalent of war, and equated the heroism of the engineer to that of ancient warriors: "It is a triumph more glorious, because far more useful to mankind, than was ever won by conqueror on the field of battle. May the Atlantic telegraph, under the blessing of heaven, prove to be a bond of perpetual peace and friendship between the kindred nations, and an instrument destined by Divine Providence to diffuse religion, civilization, liberty, and law throughout the world." The next morning a grand salute of one hundred guns resounded in New York City, the streets were decorated with flags, and the bells of the churches rung; at night the city was illuminated.[8] This exchange through an electric telegraph wire laid under the Atlantic Ocean thus marked the beginning of the virtually instantaneous communication between the continents that we e-mail and telephone users are so accustomed to today.

Laying this enormously heavy cable over thousands of miles on an ocean floor miles deep was an extraordinary and unprecedented activity. Its conception and its undertaking bear witness to the technological imagination and engineering prowess of the nineteenth century. But that the transatlantic cable failed after this exchange, not to be restored to useful service until 1866, illustrates both the difficulties and the accomplishments of nineteenth-century science and engineering.

With the success of telegraph systems within England consisting of copper wires strung on poles along the railway right of way, the notion had immediately emerged of connecting the domestic telegraph system with countries across the water. What was needed for an undersea connection was a good insulator to cover the wire and prevent the electric current from leaking into the surrounding water. In 1842, an insulating gum was found which could be melted by heat and readily applied to wire. This material was gutta-percha, the adhesive juice of a tree found in southeast Asia. In August 1850, John Watkins Brett's Anglo-French Telegraph Company laid the first line across the English Channel. It was a copper wire coated with gutta-percha but without any other protection. The next year, a cable with a more fully protected wire core was laid from a government ship which towed the heavy cable across the English Channel. In 1852, Great Britain and Ireland were linked together. In 1853, England was joined to the Netherlands by a cable under the North Sea laid by a paddle steamer which had been fitted for the work. Again technological innovations functioned as a system since it was the newly developed steamship, even in its earliest form as powered by paddle-wheels, that made possible the laying of the bulky continuous cable.

Having joined the island nation to the European continent, the next great project for the Victorians became joining England and America across the ocean. The driving force for this project was the American Cyrus Field. Like the railway system, this vast engineering project demanded a vast amount of what we would call start-up capital. And as in the case of the railways, the money was supplied through the creation of a limited-liability stock company offering corporate shares to the public. Field raised the funds in the Atlantic Telegraph Company by selling shares to both British and American investors. With great difficulty he also garnered a grant from the United States government.

In 1857, two iron steam-powered ships, one British and one American, set out to lay the cable, one sailing from the east, the other from the west with a planned splice in the middle. Cable had been laid from Washington to Newfoundland and from London to Ireland. The great unwired territory was the uninterrupted stretch of twenty-three hundred miles across the Atlantic. The cable itself consisted of seven copper wires covered with three coats of gutta-percha and wound with tarred hemp, over which a sheath of eighteen strands, each composed of seven iron wires, was laid in a close spiral. The cable weighed nearly 1.1 tons per mile. Yet within all this material, the cable was still relatively flexible. Given the engineering difficulties it was no surprise that the first attempt in 1857 failed when the cable broke. But drawing upon the lessons of that experience, a second attempt in 1858 was successful.

As is so often the case, the exploits of the engineer-as-hero, like that of the soldier, did not meet with lasting victory, but demanded further battles. The cable completed in 1858 that carried the message between queen and president soon failed. But experience led to innovation, particularly with the insulation. A new cable was designed with a copper core covered with improved insulation of gutta-percha, hemp, and soft steel, although this made the cable heavier than the original. The weight of the new cable was nearly twice the weight of the old. Just as the immense weight of the railway carriages demanded new methods of building the bridges spanning the rivers of England, the task of carrying the great weight and bulk of this enormously heavy cable for twenty-three hundred miles demanded new technologies. For laying this improved but far heavier cable, the engineers called into service the exemplar of the Victorian steam-powered iron ship, the *Great Eastern*, the largest iron ship of the time (described more fully in the section "The Iron Steamship" in this chapter). She was fitted with three iron tanks for storing the cable, and her decks furnished with the paying-out gear. Again, the first attempt failed when the cable snapped, but a second attempt succeeded in 1866. The *Great Eastern* recovered by a grappling iron the lost cable at a depth of two and a half miles and spliced it to the cable in her hold to create two working telegraph lines.

With the monumental engineering feat of the laying of the transatlantic cable achieved, an unforeseen problem appeared in this unprecedented transmission of electrical currents through long lines of wire surrounded by water. The signal on the cable began to slow and gradually disappear. Telegraph lines running on land showed no such problems. It was soon understood that since water conducts electricity, the electrical pulse in the cable sets up a charge in the surrounding water, and the interaction of the two charges blurs the signal, especially the sharp distinction between the dots and dashes of Morse code. This phenomenon increased over the long distances of the transatlantic cable.

In a manner indicating the new importance of theory to technology by the mid-century, the difficulty was eventually resolved by a scientist. William Thomson, later Lord Kelvin, was a university scientist who in the 1840s had done important theoretical work in thermodynamics, the study of heat energy. But he gradually moved from theoretical science to engineering, becoming particularly interested in the problems of the undersea cable and even sailing on the *Great Eastern* cable-laying expedition. His main contribution to cable technology was the invention of the mirror galvanometer, a device highly sensitive to small electrical charges that magnified and clarified the weak electrical signals of the cable. Basically, a small mirror is hung with small magnets on its back. A beam of light is directed at the mirror. The incoming electric charge as well as the absence of current move the magnets that in turn move the mirror, thus changing the direction of the reflected light. The shifts of light on the screen indicate the dots and dashes of Morse code and can easily be read by a trained operator. Yet, for all these innovations, by modern standards the rate of transmission was quite slow, eight words per minute.

With the success of the transatlantic cable, the nineteenth-century Internet spread from Britain to its colonies. The cable was laid from London, the metropolitan center of Empire, to the farthest colonies because of a sense of urgency about imperial control. Tendrils of wire reached through the Suez Canal to India, then under the Pacific Ocean to Malaysia, Australia, and New Zealand. Lines moved also in a southern direction under the Atlantic to the British possession of Nigeria in western Africa and from the Suez to the British colonies of east Africa. Significantly, the spread of the undersea cable enabled time to become globalized, since a time signal could be transmitted quickly and accurately through the world. With imperial England now the world's most powerful nation, Greenwich Mean Time became and continues to be the standard of time for the entire world.

THE IRON STEAMSHIP

The expansive energy of Victorian technological innovation moved communication beyond the confines of England to the entire globe. As the lines of copper wire along the railways linking the cities of England expanded into the undersea cables connecting the continents, so the nineteenth-century application of steam power to motion led to the building of huge steam-powered iron cargo ships and ocean liners that linked the continents and the far-flung possessions of the Empire.

The development of the steam-powered ship, however, was not as quick and dramatic as the explosive growth of the railway. There were technological problems inherent in ocean travel and the continuing competition of sail. As in the case of the steam engine, in the development of the steam-powered ship one cannot look to a single invention, but rather to continuous innovation that commenced late in the eighteenth century. Indeed, the application of steam to travel on lakes, rivers, and canals predated the full development of the steam locomotive running on rails.

Like the steam-powered locomotive, the steamship depended on the development over time of the reliable steam engine. The first steam-powered ships transformed piston movement generated by steam to rotary motion. The early steamboats and steamships (following convention, the term steamboats applies to vessels used on lakes and rivers, and steamships to vessels designed for ocean travel) used the engine to turn paddle-wheels located outside the hull, either at the stern or along the sides of the boat. In 1802, the *Charlotte Dundas*, commissioned by Lord Dundas of Scotland, towed two seventy-ton barges nineteen miles along the Forth and Clyde Canal to Glasgow, Scotland. This vessel is generally considered the first practical steamboat.

The first commercially successful boat powered by steam is generally credited to Robert Fulton, an American. The development of Fulton's steamboat, like the laying of the transatlantic cable organized by Field and the work of Morse on the electric telegraph, illustrates the globalization of invention and its quick spread among industrial nations in the nineteenth century. Fulton actually built his first steamboat while working in France, and sailed this small ship successfully in 1803 on the river Seine. Before returning to the Unites States, Fulton ordered a Boulton & Watt steam engine to be shipped to America. Once in America, he built at New York City docks what he called the *North River Steamboat*, more generally known in history as the *Clermont*, using the Boulton & Watt engine, the most advanced propulsive device of its day. The wooden ship was long and slender, 141 feet in length, 14 feet wide. There were paddle-wheels on each side and like most early steam-powered vessels it carried a set of sails on tall masts.

On its initial voyage on August 17, 1807, the *Clermont* traveled the 130 miles from New York to Albany on the Hudson River, inaugurating the first commercial steamboat service in the world. At an average speed of five miles per hour, the trip took thirty hours including an overnight stop, thus reducing the time between the cities from the four days required by sail. The initial voyage of what was called originally "Fulton's monster" was described in an 1807 publication with the same fear and wonder that greeted the first steam locomotive in England twenty-three years later:

> The surprise and dismay excited among the crews of these vessels by the appearance of the steamer was extreme. These simple people, the

majority of whom had heard nothing of Fulton's experiments, beheld what they supposed to be a huge monster, vomiting fire and smoke from its throat, lashing the water with its fins, and shaking the river with its roar, approaching rapidly in the very face of both wind and tide. Some threw themselves flat on the deck of their vessels, where they remained in an agony of terror until the monster had passed, while others took to their boats and made for the shore in dismay, leaving their vessels to drift helplessly down the stream. Nor was this terror confined to the sailors. The people dwelling along the shore crowded the banks to gaze upon the steamer as she passed by.[9]

With the success of the initial run, the *Clermont* began a regular passenger boat service along the Hudson that was a clear commercial success. Similar lines were established in England. Although inland travel by sail continued in America and England because of the expense of building steamboats, the era of domestic water travel by sail was clearly vanishing.

The success of steamboats on the rivers and canals of England and America inevitably led to plans for steam-powered travel over the oceans. The ocean-going steamship, however, faced a number of difficulties. For one, there was a problem in generating steam not encountered by the locomotive or even by the *Clermont* steaming on the Hudson. For its operation, the steam engine needed a large supply of fresh water for the boiler where the water was heated to generate the expanding steam. Large water tanks were set up along the railway lines to replenish the water supply for the boiler of the steam locomotive. The *Clermont* could take fresh water from the river. But there was an obvious difficulty in finding fresh water in the middle of the ocean. If sea water were used for the ship's boilers, the boilers had to be cleaned regularly to remove the accumulated salts in an expensive process that took the ship out of use and shortened the life of the boilers. It was not until the middle of the nineteenth century that the Watt engine was improved for ocean travel by the development of a surface condenser in which the original fresh water of the boiler could be recycled. In this innovation, the steam generated from the original supply of fresh water was condensed by cool sea water playing on the surface of the condenser and recycled to generate more steam, thus avoiding need for sea water in the boiler.

In spite of such innovations, steamships frequently broke down in mid-ocean for mechanical reasons, causing delay. Thus, the first steam-powered ships to cross the Atlantic employed a combination of sails and steam. In 1819, the *Savannah*, an American sailing ship, became the first vessel refashioned to cross the Atlantic Ocean employing steam power. She had steam engines and two paddle-wheels that could be folded away on deck, but used her engine only a fraction of the time. In 1831, the Canadian paddle streamer the *Royal William* crossed the Atlantic primarily under steam, although she used sail when she needed to stop her engines for twenty-four hours to scrape the accumulated salt deposits from her boilers. In 1838, the *Sirius*, a wooden paddle steamer chartered by the British and American Steam Packet Company of London, became the first ship to cross the Atlantic with the continuous use of steam power. Into the last decades of the century, steamships carried sails in case of the failure of steam.

The first steamship constructed specifically for transatlantic travel was the *Great Western*, built in England in 1837. As an extension of the Great Western Railway, this ship was conceived quite literally as a way of continuing the railway across the water. The railway line took passengers to the port of Bristol, from where the liner transported them to the United States and Canada, much as the existing Channel boat train would take passengers from London to Dover, where they boarded a cross-Channel steamboat. Demonstrating the systemic connection of the steamship and the railway, the designer of this steamship was Isambard Kingdom Brunel, the British engineer who designed the iron bridges of the Great Western Railway.

The *Great Western*, 236 feet in length, combined the traditional and the new in marine technologies. She had an oak hull and four masts for sails, but she was also powered by steam. A two-cylinder steam engine drove the paddle-wheels on each side. As in Victorian railway stations, there was a mixture of industrial and pre-industrial styles. The interior had Gothic-revival arches and elegant wall-panel paintings done in an eighteenth-century manner.

The *Great Western* made her maiden voyage from Bristol to New York in April 1838, marking the beginning of scheduled transatlantic steamship service only eight years after the opening of the first regularly scheduled railway. The transatlantic passage took fifteen days and had the regularity and reliability of the Victorian railway system. The *Great Western* made transatlantic crossings until she was broken up in 1857.

Innovation continued in the building of steamships. In 1843, Brunel built the steamship *Great Britain*, the first true ocean liner, setting the design for future ocean-going ships. At the time of her launch she was by far the largest ship in the world, over one hundred feet longer than her rivals. Rather than wood, her double hull was constructed of wrought iron with watertight bulkheads. Crucially for the history of maritime propulsion, the paddle-wheels were replaced by a sixteen-foot iron underwater screw propeller driven by the ship's steam engines. Designed initially for the transatlantic luxury passenger trade, she could carry 252 first- and second-class passengers and 130 crew.

The history of the *Great Eastern*, Brunel's successor to the *Great Britain*, is one of the most famous and most tragic stories of the Victorian engineer-as-hero. A contemporary photograph of Brunel against the massive launching chains of the *Great Eastern* captures the vast scale of the project and

Brunel's hubris, or overreaching pride. As in Greek tragedy, hubris was the very quality of genius embodied by the engineer-as-hero, the quality of "thinking big" that destroyed him and his great iron ship. (The same narrative of pride in technology was played out early in the next century in the familiar story of the sinking of the giant steamship *Titanic* on its maiden voyage in 1912.)

Isambard Kingdom Brunel on the *Great Eastern*. This 1857 photograph by Robert Howlett shows Brunel, the builder of the massive steamship the *Great Eastern*, against the launching chains of his ship while still in its dockyard. The pose exemplifies the self-confidence of the Victorian engineer as hero as well as the monumental scale of his creation. [Permission of National Railway Museum/Science & Society Picture Library.]

Built on an unprecedented scale (it was originally called the *Leviathan*), the *Great Eastern* was conceived with the intention of linking Great Britain with India, the centerpiece of Empire, via the Cape of Good Hope at the southern tip of Africa, without the stops for coal demanded by long steam voyages. Speedy, economical, nonstop voyages would be made possible by constructing a ship large enough to carry the coal for the entire voyage outwards and, if necessary, for the return voyage. The huge ship would have the capacity to carry four thousand passengers around the world without refueling, similar to the jumbo jet planes built in our own time. The ship took four years to build. At her launch in 1858, she was the largest ship ever built, seven hundred feet long, eighty-three feet wide. The double hull was constructed of 0.75 inch thick wrought-iron plates.

Although the ship carried sails on six masts, like the *Great Britain* it depended primarily on screw propulsion. The four-bladed screw propeller was twenty-four feet across, an unprecedented size, and powered by engines built by Boulton & Watt. These engines weighed five hundred tons and drove a thirty-six-ton cast-iron propeller via a shaft 150 feet long and weighing sixty tons. There were two paddle-wheels fifty-six feet in diameter for backup and use in shallow water. These had their own dedicated steam engines.

The story of the ship is one of difficulties generated by the ship's massive size. After accidents and delays in launching, the ship finally had its sea trials in 1859. But this was to be Brunel's final great project, as he died four days after these trials. His death was probably due to the physical fatigue and mental strain of superintending the building of this unprecedented ship. The ship was a failure. Because of her size, the ship was slower than her rivals and rolled more in heavy seas. Few passengers signed up. The first transatlantic voyage in 1860 was not profitable. After a series of accidents and high costs, the vessel was removed from passenger service.

After her huge size prevented her from being a profitable passenger liner, the *Great Eastern* was given a new and productive life as a cable-laying ship. Some boilers were removed as well as great parts of the passenger rooms and public spaces to give way to open-top tanks for storing the coiled cable. Not only did she lay the 1865 transatlantic telegraph cable, but also from 1866 to 1878 laid submarine telegraph cable from Europe to Aden on the Arabian peninsula and then on to Bombay, India. She was finally broken up for scrap at Liverpool in 1889–1890. It took eighteen months to take her apart.

As the nineteenth century drew on, in the area of steam-powered ocean travel, as in other realms, the process of continuous innovation accelerated. Speed improved; the time of a transatlantic passage was reduced from fifteen days at mid-century to six days by the end of the century. As wooden hulls had been replaced by iron, so iron was replaced by the stronger material of steel. The paddle-wheel was universally replaced by the single screw propeller. By the beginning of the twentieth century, twin screws ensured that sails no longer were necessary as backup for a malfunctioning propeller. Also, by the end of the nineteenth century, the propellers were powered by highly efficient steam turbines.

In social terms, the development of large, reliable steamships for intercontinental travel had enormous global effects, particularly in enabling the mass migrations of populations that marked the nineteenth century. For Victorian England, the availability of inexpensive if greatly uncomfortable steamship travel in steerage allowed the large-scale emigration of the increased population that accompanied industrialization. The steamship enabled British men and women to seek new opportunities abroad, especially in the white settler colonies of Canada, Australia, and New Zealand. This safe and now-cheap travel also made possible the great waves of immigration that peopled the growing United States. And the steamship held together the global British Empire.

THE ROMANCE OF CABLE AND STEAMSHIP

One of Victorian England's favorite paintings was Joseph Mallord William Turner's *The Fighting Temeraire tugged to her Last Berth to be broken up* of 1838.[10] This painting shows the wooden warship that had fought against France towed off to the wrecking yard by a prosaic steam-powered tug. The scene juxtaposes the heroic age of naval battle with wood and sail to the commercial values of an age of iron and steam, thus manifesting one of the central themes of the Victorian culture of technology, old and new technologies both existing in the same moment, yet one doomed to extinction. Turner captures here the mingling of nostalgia and progressive vision in the Victorian imagination. In Turner's *Rain, Steam and Speed*, adulation of the power and energy of the railway replaced the romance of the stagecoach. By the end of the century, an aura of beauty came to clothe the ocean-going steamship and an air of mystery the undersea cable that had wired the world.

For the celebrated British poet of empire, Rudyard Kipling, writing in the last decades of the century, the submarine cables that bind the empire become live creatures within the unseen world of the deep oceans. In his poem, "The Deep-Sea Cables," the undersea wires become "hands" joining the far-flung bodies or colonial possessions of England. They become animate creatures as they "creep" slowly across the ocean floor. The message they transmit is Kipling's imperial dream of global unity as "Men talk today o'er the waste of the ultimate slime,/ And a new Word runs between: whispering, 'Let us be one!' "[11]

The ocean liner, too, manifested its own particular splendor as again evoked by Kipling. Born in India, the literary spokesman of British imperialism quite rightly sees the steamship, like the undersea cable, as tying the empire together. A bard of the Victorian cult of the engineer, in his great celebratory poem of the machine, "McAndrew's Hymn" (1893),[12] Kipling establishes as hero a "dour Scots engineer." The choice reflects the prominence of Scotsmen among Victorian technologists, going back to James Watt, and forward to Scotty, the engineer of the Starship *Enterprise*.

McAndrew's reply to an upper-class visitor to his engine room who asks whether "steam spoils romance at sea" is that there is need for a romance of technology and a new poet "like Robbie Burns to sing the Song o' Steam!" For the sensibility of Kipling's engineer, as the precisely engineered parts of the marine engines transform the energy of steam into mechanical motion they create this "Song o' Steam" equal to the compelling harmony of a symphony orchestra as "the crank-throws give the double-bass, the feed-pump sobs an' heaves,/ An' now the main eccentrics start their quarrel on the sheaves."

As the engine manifests the fresh beauty of a harmony of metal parts in motion, so the heroism of a ship's engineer equals the derring-do of the pirate or the commander of a wooden man-of-war engaging an enemy ship. Like that of the marine engine, the engineer's behavior is regular and reliable, unobserved yet crucial to the knitting together of the colonies. The engineer carries out what appeared to Kipling and to the British the divine mission of empire as he oversees the travel of "average fifteen hunder souls safe-borne fra' port to port."

IRONCLADS, MACHINE GUNS, AND EMPIRE

The globalization of transport and communication technologies through the transformation of the steam-powered locomotive and the electric telegraph into the ocean-going steamship and the undersea cable was crucial to the expansion and maintenance of the British Empire. The relation of invention to empire presents a complex instance of the question of technological determinism. Does technological change have a built-in dynamic so that the generation of one invention by another is inevitable as determined by the very nature of technology? Or, conversely, do social and cultural values encourage as well as limit invention? Does innovation control and determine the shape of society or does society and its values determine the pace and nature of technological change? In economic terms, does technological innovation shape the economy or do economic needs generate new technologies? In this case, did the expansion of the British Empire in the nineteenth century generate the need for the technologies of the factory system of production, steamships, and undersea cables? Or did the technologies, particularly the highly productive factory system, generate the expansion of empire?

By the end of the nineteenth century, the British Empire covered the globe. In the popular phrase of the time, "The sun never sets on the British Empire." The empire included the colonies of Canada, Australia, and New Zealand where British people settled on what was imagined to be empty land, though of course it was occupied by indigenous people who were then marginalized, displaced, or exterminated. India was considered the "jewel in the crown" of Victoria's rule, held under firm military control after a rebellion at mid-century. In what is known as the "Scramble for Africa" among European nations in the latter decades of the nineteenth century, such possessions as Egypt, Kenya, Uganda, Nigeria, and the Sudan came under British control. After bloody late-century wars with the native Zulus and the Dutch settlers (or Boers), what is now known as South Africa was added to the empire.

There is a strong and in many ways compelling argument that this expansion of British domination over areas of Africa and Asia in the Victorian age, although motivated in some degree by the drive for strategic power and the missionary effort of conversion to Christianity, was driven primarily by the economic needs created by the factory system. Furthermore, the need to control these millions of subject people generated or quickened innovation especially relating to the steamship, the undersea cable, and the machinery of warfare.

Consider first the question of raw materials. In the pre-industrial world, the supply of wool from the sheep grazing in the fields of England was sufficient for the scale of production by the domestic spinning wheel and the hand loom. With the emergence of the steam-powered textile mill, the ability to produce woven textiles with speed and in great quantities demanded raw wool and cotton in huge quantities to match the productive capacity of the spinning jenny and the power loom. Thus, England had to look abroad for its raw material. To supply this need, much of Australia and New Zealand was transformed into sheep ranches. The development of fast and reliable steam-powered cargo ships allowed the wool to be shipped thousands of miles to the Midlands mills where the steam-powered machinery could transform the fibers into finished textiles.

The political effects of the factory system appear in the struggle of nineteenth-century British mills to ensure a reliable supply of cotton. Since cotton could not grow in England, all the material had to come from abroad. Although the possessions of India and Egypt supplied some of the fiber demanded by the mills, the main source of cotton for England was the American South, where until the end of the Civil War it was grown on plantations worked by slaves. Although there was some competition from fast clipper ships under sail, steam-powered freighters carried the bales of cotton from the American South to such English ports as Liverpool, from where the raw material was transported by rail to the Midlands mills.

This dependence of Victorian textile mills on American cotton created a major diplomatic crisis during the American Civil War in the 1860s. The blockade of Confederate ports by Union ships created in England what was called the cotton famine, a shortage of cotton that closed the mills, created severe unemployment, and led to a serious loss of capital. This shortage generated a dilemma for the British government. Should the British try to break the blockade, even at the risk of going to war with the Union, in order to relieve the mill-workers of the Midlands? Or should Britain support the Union in what was generally seen as a moral crusade against slavery? That the workers supported the Union side even at the cost of their income can be seen in an "Address from the Working People of Manchester to His Excellency Abraham Lincoln, President of the United States of America" in 1862: "The vast progress which you have made in the short space of twenty months fills us with hope that every stain on your freedom will shortly be removed, and that the erasure of that foul blot on civilisation and Christianity—chattel slavery—during your presidency, will cause the name of Abraham Lincoln to be honoured and revered by posterity. We are certain that such a glorious consummation will cement Great Britain and the United States in close and enduring regards."[13] Naval confrontation between the United States and England was averted by the victory of the Union which allowed cotton shipments to England to resume. But the potential for a devastating war between nations over the crucial need for raw materials to feed an industrial system suggests the economic motives of empire and the danger of competition among industrialized nations over the raw materials, such as cotton in the nineteenth century and oil in our own time. The vastly increased scale of mechanized production, then, generated a struggle for imported raw materials that was resolved by imperial expansion and control.

Furthermore, just as nineteenth-century England could not provide the wool and certainly not the cotton for its mills, so England could not provide enough consumers for the virtually limitless stream of manufactured textiles, not to speak of the domestic manufactures shown at the Crystal Palace. Buyers were also needed for the machine tools as well as heavy machinery, such as locomotives, railway cars, steam hammers, and steamboats made in England. In many ways, the machine made not only textiles, but also made the British Empire. The zeal of England and of other European nations for colonies in the industrializing nineteenth century can be seen in part as a desperate attempt to gain quite literally a captive population to purchase the flood of manufactured goods pouring from the factories and foundries.

Cotton and wool were shipped by steam to England where they were transformed into wool and cotton textiles. The finished textiles were then sent back by steam-powered cargo ships to the empire as clothing and blankets and curtains to be sold to the colonial populations. Millions of buyers in an India ruled by Britain bought saris made from cotton grown in India and transported by steam to the English Midlands. As early as 1830, more than half the value of British home-produced exports consisted of cotton textiles. Rather than shipping bales of wool and cotton to England to be made into cloth, it might have been more practical, even more economical, to set up mechanized mills in Australia, New Zealand, and India to manufacture clothing. The finished products could then have been sold locally and the surplus shipped to markets of England and Europe. But here we must note the political reality of British imperial control which subordinated economic efficiency to the welfare of the workers and owners within the textile industry of the mother country. British rule ensured that the colonies functioned only as exporters of raw material to the factories of England and prevented the colonial territories from becoming competitive producers of finished goods.

The rise of the machine in the nineteenth century created the economic globalization under which we still live. Within the Victorian colonial system, one of the main consequences of mechanized production was the division of the globe into two types of nations—the suppliers of raw material and the exporters of machine-made goods, or in modern parlance, the undeveloped and the developed countries. The prosperity of factory workers, factory owners, and traders within an industrial Victorian England was supported by a system of colonial suppliers and markets. Colonial control was maintained by the machine technologies of rapid communication through the undersea telegraph, the swift and reliable transport of the steamship, and ultimately on military power based on innovative machines of war.

EMPIRE AND MACHINES OF WAR

While this volume has so far focused on the rise of the machine in factory production, transportation, and communication, an accurate account of technology in the nineteenth century must include the mechanization and industrialization of warfare. Throughout the Victorian period, England was engaged in a multitude of small colonial wars in such areas as the Sudan, Afghanistan, and what is now South Africa. Victory was in many cases achieved through innovations in weaponry. If the machine shrank the globe into a tight structure allowing for the flow of raw materials to the center and the return of manufactured goods to the periphery, this imperial system was ultimately sustained by the threat and by the use of overwhelming military force based on technological supremacy. The imperial conquests of the nineteenth century, especially in the scramble of European nations including England for territory in Africa in the latter decades of the century as well as the continued control of the Indian subcontinent, for all the rationalizations about extending the benefits of civilization, were grounded on the efficiency of a mechanized army and navy.

Throughout the nineteenth century, the military was at the forefront of innovation. In the domain of war as in domestic manufacture, the British Army and the Royal Navy had the same demands as civilian consumers for large quantities of goods, chiefly guns and ships, that were reliable and that could be produced quickly and bought cheaply. To achieve these goals, the principles of the machine revolution were deployed to military production: shifting from human to mechanical power; standardization of components, particularly in the use of interchangeable parts; concentration of production in large workshops; and disciplining of the workforce to perform routine mechanized activities.

The most famous and the most historically important application of the machine to military production in nineteenth-century Britain was the shift to mechanized, assembly-line manufacture of pulley blocks for wooden sailing ships at the Portsmouth Dockyard of the Royal Navy. Pulley blocks are hollowed-out wooden blocks with a rotating grooved wheel inside through which the ropes are strung that raised and lowered the many sails of a man-of-war. These components were crucial to the Royal Navy. In the first decades of the nineteenth century, with the Royal Navy still powered by sail, as many as 922 of these blocks were needed on a single man-of-war. The fleet as a whole needed one hundred thousand pulley blocks. Before the change to the mechanized production line some simple machines were used. Yet, a skilled workman using hand tools was needed to complete most of the work. With the demands of the war with France in the first decades of the nineteenth century, the craft process simply could not supply the demand for this crucial element of a fighting ship.

In 1801, Marc Isambard Brunel (father of Isambard Kingdom Brunel who engineered the railway bridges of the Great Western Railway and built the *Great Eastern* steamship), approached the dockyard with a scheme for making blocks with a set of special machines designed by Henry Maudslay, one of the most famous of British mechanists. In 1802, Samuel Bentham, the Inspector General of Naval Works, approved the installation of the machinery at Portsmouth Block Mills. It is indicative of the spirit of the age that Samuel Bentham was the brother of Jeremy Bentham, the influential proponent of the Utilitarian philosophy that saw moral choice as a calculation of the pleasure and pain resulting from an action.

As one of the first instances of the complete mechanization of a manufacturing process, Brunel's scheme was an important step in the history of machine production. The dockyard established one of the first production lines consisting of a set of single-purpose machines. Operations were performed by machines built to execute only a specific function. The number of each type of machine was determined in proportion to time taken to perform the operation so that there was a steady flow of production. There were forty-five machines in a line, with the machines made in three sets to make a range of different-sized blocks. There were machines for sawing a log into blocks, for boring the interior of the wooden shell, and for cutting the exterior of the block into the proper shape. Single-purpose machines cut the rough form of the pulley discs from wood, shaped the sheaves or pulleys, implanted a bronze fitting, and drilled a hole for the axle. Under the old system of hand production, the blocks within each category of size had slight differences in their dimensions. Crucially, the machine-made blocks within each size category were standardized and thus interchangeable. This accuracy was achieved through building the machines entirely of iron, since iron did not wear with use. The machines were highly accurate and durable. Some were still in operation as late as the mid-twentieth century.

By 1805, machinery took over production of the navy's entire block supply. With the shift to the mechanized assembly line, the Royal Navy solved its supply problem in the midst of war with France. The rate of production increased tenfold and by 1808 the plant was producing 130,000 blocks per year.[14] As in the shift from hand to machine methods in the textile mills, mechanized manufacture was far less labor intensive, no longer requiring skilled wood-carvers, but only unskilled workers to function as machine watchers. Ten unskilled men were able to equal the output of one hundred block makers. The capital cost of the project was recovered in three years. With the nineteenth-century interest in innovation as spectacle that was continued by the crowds at the Crystal Palace, the dockyard itself became a destination for tourism devoted to visiting industrial sites. In 1805, Marc Brunel urged Samuel Bentham to erect a fence around the block mill to keep visitors out.

With its creation of interchangeable parts, mass production perfectly suited the manufacture of weapons. The production of arms, once located in small workshops that had forged swords and armor, was in the nineteenth century relocated to armories that were essentially factories for manufacturing military equipment. As in the cases of the steamboat and the telegraph, there was a transatlantic cross-fertilization of ideas and techniques. The central figure here is the American, Eli Whitney. Whitney is best known for his invention of the cotton gin at the end of the eighteenth century. This was a mechanical device that efficiently removed the seeds from the long-staple cotton grown in the American South. This machine transformed southern agriculture by making possible the shipping of large quantities of cotton as bales of seedless cotton to the textile mills of England. As much as the spinning jenny and the power loom, Whitney's invention was foundational to the English cotton industry.

Whitney then turned his attention to making weapons on an industrial model. Drawing upon his knowledge of European methods of machine production, Whitney established a factory in New Haven, Connecticut, in 1798 to manufacture muskets for the American army. The historical importance of the factory lies in its employing machine tools to create standardized, interchangeable parts for guns. Whitney's machines were used to cut and grind the metal, to bore the barrels, and to cut wood into standardized stocks. The parts—triggers, barrels, stocks—were then assembled into the finished weapon. Like the Portsmouth mechanization of producing pulley blocks, this process was quick and efficient, supplying muskets cheaply to the American Army. Furthermore, the replacement of hand labor, with its inevitable vagaries and imprecision, created a more reliable and effective weapon, particularly since the barrels could be rifled, that is grooved, by machinery with great accuracy. The mass production of standardized parts for weaponry, what came to be called the armory method, was quickly adopted in England, particularly at the Woolwich Armory that became the major producer of weapons for the British Army.

The application of the machine brought industrialization not only to the armory, but also to the battlefield. Fighting itself became industrialized. As in the factory, so on the battlefield, mechanization brought on a deskilling of the operative. The soldier no longer had to perform dexterously with sword or pike. Innovation in weapon technology had done away with the muzzle-loading musket. Now, given a reliable machine-made weapon, the soldier had only to pull the trigger, a simple repetitive act analogous to the work of a machine tender. Armies now focused on training in close-order drill rather than individual heroic action. Soldiers were disciplined to function as a single unit on command. They were drilled to carry out orders to charge, to defend in set patterns such as the square, and to fire in unison or in sequence. In short, the army had become a machine, a military machine that, like a factory, forged human beings and mechanical devices into a single entity of disciplined activity.

Mechanization also brought to the operative either as worker or soldier an enormous amplification of force. Here the exemplary invention is the machine gun whose name nicely symbolizes the connection between the rise of the machine and the increasing power of weaponry. The machine gun or, more accurately, the automatic weapon, exemplifies the continuing application of the idea of the automatic to create a machine working without the need of human intervention. Indeed, guns that repeat firing are still called automatics. In the rifle of the mid-nineteenth century, a soldier had to load the cartridge by hand into the breech of the gun, then fire each round by pulling the trigger. The trigger then moved a pin that hit the cartridge containing the explosive gunpowder which propelled the bullet out the barrel. In the machine gun, the work of loading and firing is done with great rapidity by the machine gun itself. As in the working of the power loom, individual human action disappears once the machine is started.

What is generally considered the first machine gun was invented by an American, Richard Gatling, in 1862, during the American Civil War. The Gatling gun, as it was called, has six to ten barrels within a single cylinder. The operator turns a crank which rotates the barrels within the cylinder. The movement allows a firing pin to hit the bullet; the spent shell is ejected by the explosion; and from the storage carousel of bullets a new cartridge falls into place. Thus, a single soldier can achieve the deadly effect of rapid fire by simply turning a crank that continues an automated process.

Further innovations for the automatic weapon were made by Hiram Maxim, another American. Invented in 1881, the Maxim gun used the energy of each bullet's recoil force to eject the spent cartridge and insert the next bullet. The gun, like textile machinery, was thus self-acting, needing no human participation in the process. With this method the Maxim gun could fire five hundred rounds per minute, the firepower of about one hundred rifles. The Maxim gun was also the first water-cooled machine gun. It used a water jacket around the barrel that tended to heat rapidly because of the quick firing. This portable water-cooled machine gun was adopted by the British Army in 1889.

The killing power of the fully automated gun that efficiently applied the principle of the machine to firing bullets reached its full potential for slaughter in World War I. As troops went "over the top," charging from their trenches across open fields, they were mowed down like wheat under a machine combine, slaughtered in the thousands or tens of thousands. On one day in 1916, in the Battle of the Somme in France, over fifty-eight thousand British soldiers were killed and wounded, primarily by German Maxim guns.

In the nineteenth century, the power of the machine gun was central to the colonial wars Britain fought to capture and control territory. The Maxim gun was particularly lethal in the colonial wars in Africa and thus a major factor in Britain's success in securing African territory for the empire. In one battle in the war fought in South Africa against the Ndebele nation in 1893–1894, an army of seventeen hundred Ndebele attacked a camp defended by fifty Europeans. In their traditional way of warfare, the Ndebele courageously charged across open ground, much like the British at the Somme, only to be cut down by four Maxim guns. The effect of these guns against Africans, as against the Europeans in the twentieth century, was as much psychological as physical. The Ndebele army broke and ran.

In a popular phrase, England before the rise of the machine was said to be defended by wooden walls, that is, by the wooden men-of-war that patrolled the seas around the island nation and protected colonies abroad from other nations. In the nineteenth century, these wooden walls were replaced by iron and eventually by steel hulls. Sails were replaced by steam-powered screw propellers. Within the domain of warfare, the move from wood to iron and from sail to steam brought into being the dominant weapon of naval warfare and imperial control, the steam-powered ironclad, as the nineteenth century termed a fighting ship fully protected by metal armor.

The inventions that brought into being the swift ocean liners also created the ironclad. Advances in metallurgy allowed the casting of strong steel and iron for ships' armor. Improved machine tools enabled the precise boring and rifling of barrels for accurate long-range guns on board the ship. Innovations in marine engines gave rise to warships with long cruising range. The replacement of paddle-wheels by screw propeller enabled swift battleships and cruisers.

England moved only reluctantly from its proud tradition of wooden men-of-war under sail. The change by the Royal Navy to the fully developed, steam-powered, fully armored warship was uneven. France and England had built wooden-hulled, steam-powered screw-propeller frigates in the 1840s. This was followed by the innovation of sheathing the wooden hulls with iron plates. Only in 1860, threatened by an iron-hulled warship built by France, did the British Navy begin building a warship on a new design incorporating previous developments, the HMS *Warrior*. This ocean-going warship was powered by steam engines that could drive the ship at the then-rapid speed of 14.5 knots. It had a completely iron hull, with armor plating consisting of layers of wrought iron and teak in a belt around the sides of the ship. Older muzzle-loading guns were replaced by breech-loading guns which did not have to be moved back after each firing. The use of a screw propeller gave the ship great maneuverability.

The *Warrior*, however, did not see battle. The first battle between ironclads took place in America in 1862 during the Civil War. In order to break the Union blockade that had stopped cotton shipments to England, the Confederacy built an ironclad, the *Virginia*, better known as the *Merrimack*, to destroy the wooden sailing ships controlling the harbors of Virginia. On a cut-down hull, Confederate engineers built a slanted structure covered with four layers of iron sheets two inches thick. There were ten cannon, four on each side, one at the bow and one at the stern, as well as an iron battering ram at the prow. As slow and as clumsy as this ironclad was, on its first sortie it destroyed by ramming and gunfire two major wooden Union ships whose guns were powerless before the *Merrimack's* armor.

When the *Merrimack* steamed out the next day to complete the destruction of the Union fleet and end the blockade, it was met by the Union ironclad, the *Monitor*. This ironclad, often described as a cheesebox on a raft, consisted of a revolving tower on a very low hull. The rotating turret was armored with two inches of steel and held two large naval guns rifled for accuracy. In this first naval battle of steam-powered iron warships, each shelled the other, but could not inflict decisive damage on armored sides. Each retreated at the end of the day. Because of damage to each ship, the encounter was not repeated.

The wooden men-of-war of the Union had been saved, but the wooden warship under sail was not. The new technology of naval warfare had dramatically made the wooden ship-of-the-line obsolete. The power and invulnerability of the ironclads, as shown in the battle between the *Monitor* and *Merrimack* in which neither ship could sink the other, established in the nineteenth century an era of mutual deterrence in naval warfare between the major powers. Much like the missile race between Russia and America in the twentieth century, in the nineteenth century there was a naval arms race characterized by continuous innovation, particularly in the competition between armor plating and the power of naval gunnery. Also, the torpedo replaced the tactic of ramming employed by the *Merrimack*. This intense rivalry between the nations of Europe, Russia, Japan, and Latin America to build the dreadnoughts, as these heavily armored and heavily gunned battleships were called, continued throughout the decades before World War I.

For Britain, the technology of naval warfare also took a specialized turn in the service of maintaining the empire. England developed the gunboat, a class of smaller ironclads, steam-powered armored ships of shallow draft. The depth of a vessel's keel below the water line was usually only ten to fifteen feet. This connection of Empire and ironclad is nicely summed up in the phrase that came into being in the nineteenth century and is still in common use, "gunboat diplomacy." Even as weaponry has evolved, the term signifies the display of overwhelming military power to force compliance upon a weaker nation. The advantage and utility of gunboats in the nineteenth century was that they could navigate rivers and thus in the age before airplanes could project military power into the interior of the nation. In the Victorian era, the appearance of armored vessels in the waters of a pre-industrial nation's cities demonstrated military might in the ability to destroy the area from afar without any means of response or defense. Thus the weaker nation had to bow to the demands of the British Empire or be destroyed.

The use of gunboat diplomacy is exemplified in the actions of Britain during the First Opium War between the British and the Chinese in 1839–1842, in the pre-ironclad period. The British had a monopoly on selling opium in its trade with China. However, in 1839 the Emperor of China banned the import of opium into his country, thus destroying the lucrative British trade. The British refused to obey this edict and resorted to overwhelming force. Warships bombarded the cities of the coast; a contingent of British and Indian soldiers used the new repeating rifles against the Chinese; and armed boats sailed up the Yangtze River to destroy the tax-collecting barges of the emperor.

5

Social Effects of Industrialization: Protest and Reform

CONDITIONS OF EARLY-VICTORIAN WORKING-CLASS LIFE

In 1842, Friedrich Engels, the son of a German textile manufacturer, was sent to Manchester by his father to work in the English branch of the family business. His account of Manchester, published in 1845 as *The Condition of the Working Class in England*, provides a vivid firsthand account of the transformations brought by industrialization. This still-startling account of daily life in the early-Victorian industrial city of Manchester, what Engels termed "the classic home of English industry ... the masterpiece of the Industrial Revolution,"[1] shows clearly the social effects of the rise of the machine. Manchester of the 1840s exemplifies the sudden mass migration from the country to the industrial city, the emergence of the entirely new class of urban industrial workers, and the appalling conditions of life for these workers both within and outside the mills.

Pre-industrial England was a land of farms and sheep pastures dotted with market towns. The weaving of textiles was done in rural areas by hand-loom weavers often on their own farms. Even the one large city in England, London, was a center of government and of trade, rather than a manufacturing center. But the development of textile machinery late in the eighteenth century and further innovations early in the nineteenth century dictated the centralization of manufacturing and the need for the concentration of workers. With the factory system came the new phenomenon of the industrial city, with the largest concentration of manufacturing towns located in the Midlands of England. Into these towns flowed the population of the countryside. To nearly destitute agricultural laborers, the low wages of the mills appeared to be good wages paid regularly. The routinized indoor labor of the mills offered a release from the ceaseless outdoor drudgery of farm life. Then, as now, there were other lures of the city for the villager. The life of the city appeared to offer a personal freedom from the social constraints and boredom of village life. From Ireland, then a part of England, came an impoverished populace to find the jobs wholly lacking in their own country. The urbanization of central England was rapid and extensive. Small Midlands market towns like Manchester, Sheffield, Birmingham, and Leeds were suddenly transformed into large industrial cities. In the first half of the nineteenth century, Manchester's population more than tripled. In 1811, there were approximately 89,000 inhabitants; in 1841, over 235,000; and by 1851, more than 303,000.

For the new inhabitants of the new city, the primary problem was housing. In Manchester, the exemplary industrial city, the location of housing was generated by the demands of factory work and shaped by class divisions. The mills were at the center of the city. In an urban configuration that still can be seen in many nineteenth-century American and British cities, the mill owners lived in newly-built, spacious houses surrounded by gardens in the suburbs far away from the polluted air of the central town. Since the owners could afford carriages to take them to the mills they operated, they were able to build and drive down avenues dedicated to their swift carriage ride, sparing them the sight of workers' housing.

Since the workers had to walk to work, the workers' housing was densely concentrated around the mills at the core of the city. Housing for the newly-arrived workers consisted either of older buildings converted to hold many more people or newly built homes often of shoddy construction erected by the factory owners for their employees. Without government regulation, the only criteria for housing were density and cheapness. The modern counterparts of these Victorian industrial cities are the vast slums in such urban conglomerates as Mexico City, Mumbai, India, or Lagos, Nigeria. Then as now, with the move from an agricultural to an industrial economy, population flowed from countryside to a wholly unprepared urban space. In the Victorian industrial city, houses were crushed together to form narrow alleys and twisted lanes. The dwelling spaces were tiny. Basements provided some measure of defense against the dampness of the ground, but these were soon turned into cellar apartments. One dwelling could be shared by as many as twelve people who worked in shifts and shared the beds. There was no privacy.

Engels reports of Manchester:

One walks along a very rough path on the river bank, in between clothes-posts and washing lines to reach a chaotic group of little, one-storied one-roomed cabins. Most of them have earth floors, and working, living and sleeping all take place in the one room. In such a hole, barely six feet long and five feet wide, I saw two beds—and what beds and bedding!—which filled the room, except for the fireplace and the doorstep ... in front of the doors filth and garbage abounded.[2]

The sanitary facilities ranged from the nonexistent to the appalling. In early-Victorian Manchester, up to one hundred houses shared a privy or outhouse, usually a hole dug in the corner of a yard. There were no municipal sewage lines until later in the century. Through the 1840s, the working-class districts had no government water supply to individual houses. Workers took their water from a communal pump shared by many.

Without government regulation, the soot poured from the stationary steam engines of the mills so that a cloud of coal smoke hung in the air. Acid rain fell from the clouds during storms. Toxic waste flowed from the mills into the river Irk that ran through Manchester and provided drinking water to the communal pumps of the working-class districts. Engels provides an appalling firsthand account of the destruction of nature by industry:

At the bottom, the Irk flows, or rather, stagnates. It is a narrow, coal-black, stinking river full of filth and rubbish which it deposits on the more low-lying right bank. In dry weather this bank presents the spectacle of a series of the most revolting blackish-green puddles of slime from the depths of which bubbles of miasmatic gases constantly rise and create a stench which is unbearable even to those standing on the bridge forty of fifty feet above the level of the water.[3]

With air pollution, execrable sanitation, bad drinking water, and crowding into dank dwellings, disease spread rapidly and often. Epidemics of infectious diseases such as typhoid, spread by contaminated drinking water, came frequently. Tuberculosis spread in the teeming houses and basements. Infant mortality was extremely high in large part because of the diarrhea caused by bad water. Due mainly to disease, the life expectancy of the city worker was low. In the early industrial decades, the average age of death for a worker in Manchester was seventeen; in Leeds, another Midlands industrial town, nineteen; and in Liverpool, fifteen. As a comparison, the average age of death for a professional man in those cities was, respectively thirty-eight, thirty-four, and thirty-five.

Within the mills, too, life was dangerous for the men, women, and the children who worked there. Early-Victorian manufacturing depended on child labor. In 1830, there were over 560 cotton mills in Lancashire, the Midlands district that included Manchester. These mills employed more than 110,000 workers, of which 35,000 were children, some as young as five years of age. Children worked as long as fourteen hours in the mills, but many worked even longer since, with no government regulation, mill owners could demand any terms of employment they wanted.

This use of child labor made sense to early-Victorian workers as well as to owners. Quite simply, it was good economics for both classes. The workers had large families and the children's wages were necessary to provide for these families. Furthermore, schooling was not sacrificed since a public primary school education was not available in England in the early nineteenth century. For the owners, wages for children were very low, one-tenth of the wages paid to adult men and women. With increasing automation of textile manufacture, cheap child labor could tend many types of machinery. With children operating simple textile machines, the number of well-paying adult jobs for operating machinery decreased. Adults increasingly were used in a supervisory role. Children worked in other industrial occupations, most notably the mines, where both boys and girls began work at five years of age. Here their small size enabled them to pull carts of coal or iron ore on rails through the narrow tunnels.

Of the many female factory workers, more than one-third were married. Then, as now, women with children often had to do what was in effect a double shift of both hours in the mill and hours at home doing domestic work and child care. In the mills, because of concerns about modesty and physical strength, plus a low opinion of their abilities by the managers, women were generally kept to certain tasks resembling pre-industrial domestic activities. Continuing traditional female occupations into the machine age, women were heavily concentrated in the preparation of cotton for spinning in the hot-and-dusty card-room. Because the skilled work of operating complex machinery was seen as a male activity, women were allowed only into the simpler forms of mechanized weaving. Men were given the higher-paying tasks of operating complicated weaving machines. In the close quarters of the mill, women were subject to sexual harassment, but also achieved a certain degree of independence as they gossiped and joked with men in a camaraderie not possible in the isolation of the home.

For women, children, and men alike, the textile mill provided some degree of financial and personal freedom, but was also a place of illness induced by working conditions. Fluff, the small particles of cotton thrown off by carding, spinning, and weaving, filled the air of the mill. Women working in the card-room were especially vulnerable to byssinosis or brown lung, a disease caused by this free-floating cotton dust that destroyed lung tissue and eventually resulted in respiratory failure. Tuberculosis spread rapidly through the crowded, poorly-ventilated mills. All cotton workers suffered in the winter from the transition from the chill of the outside world to the overheated workplaces which were made artificially damp in some weaving areas to ease the process.

Working the machines in the unregulated mills also posed the constant threat of physical maiming and even death. Men and women operated spinning machines and power looms that lacked protective barriers to shield the worker, because such barriers cost the mill owner money to install. Operating machines that were not fenced in, workers frequently had hands and arms torn off. An extreme example of mechanized risk occurred when children were used to make up for the imperfections of automatic textile machinery. Small and agile, children of six and seven were employed to crawl beneath whirling looms to gather up loose cotton. These scavengers, as they were known, were particularly vulnerable to being caught up in the machinery. The industrial novel of 1840 by Frances Trollope, *Michael Armstrong: Factory Boy*, describes this work:

> A little girl of about seven years old, whose office as *scavenger* was to collect incessantly, from the machinery and from the floor the flying fragments of cotton that might impede the work. In the performance of this duty, the child was obliged from time to time to stretch itself with sudden quickness on the ground, while the hissing machinery passed over her; and when this was skillfully done, and the head, body, and outstretched limbs carefully glued to the floor, the steady-moving, but threatening mass, may pass and repass over the dizzy head and trembling body without touching it. But accidents frequently occur; and many are the flaxen locks rudely torn from infant heads in the process.[4]

Then, too, the physical conditions of labor in textile mills brought on a general deterioration of the body. Long hours in overheated rooms performing repetitive tasks without fresh air and rest for the muscles created what seemed to contemporary observers a new species of human beings. Those children who survived to adulthood had permanent stoops or were crippled from the prolonged crouching that their job required. A middle-class visitor to the mills in 1833 described the workers as deformed in body and mind:

> Their complexion is sallow and pallid. Their stature low—the average height of four hundred men, measured at different times, and different places, being five feet six inches. Their limbs slender, and playing badly and ungracefully. A very general bowing of the legs. Great numbers of girls and women walking lamely or awkwardly, with raised chests and spinal flexures. Nearly all have flat feet, accompanied with a down-tread, differing very widely from the elasticity of action in the foot and ankle, attendant upon perfect formation. A spiritless and dejected air.[5]

In 1832 a Parliamentary Commission investigating the conditions of labor in textile mills heard from a worker about such conditions:

> At the top to the spindle there is a fly goes across, and the child takes hold of the fly by the ball of his left hand, and he throws the left shoulder up and the right knee inward; he has the thread to get with the right hand, and he has to stoop his head down to see what he is doing; they throw the right knee inward in that way, and all the children I have seen, [have] that bend in the right knee. I knew a family, the whole of whom were bent outwards as a family complaint, and one of those boys was sent to a worsted-mill, and first he became straight in his right knee, and then he became crooked in it the other way.[6]

For this dangerous and often-crippling work, the early industrial workers were paid a wage that was barely at the subsistence level. The first industrial generation was deeply puzzled that in the 1840s, known as the Hungry Forties, the rise of the machine had created vast wealth in material goods and yet had generated widespread poverty, even starvation among workers in the industrial towns. Writing in 1843, Thomas Carlyle, a Victorian critic of industrialization, likened the seeming paradox to the legend of King Midas. Midas was given the gift of turning all objects he touched to gold. But when he touched his daughter, she turned to gold and died. His touch also turned his food to gold and he died, like unemployed mill-workers, of hunger. To Carlyle, like the golden touch of King Midas, industrialization had created enormous wealth, but also the death of children and the starvation of many:

> Our successful industry is hitherto unsuccessful; a strange success if we stop here! In the midst of plethoric plenty, the people perish; with gold walls, and full barns, no man feels himself safe or satisfied. Workers, Master Workers, Unworkers, all men, come to a pause.... Midas longed for gold.... He got gold, so that whatsoever he touched became gold,—and he ... was little the better for it. Midas had misjudged the celestial music-tones.... What a truth in those old fables![7]

This seeming contradiction of industrialization, known in the time as the "Condition of England Question," has several explanations, all related to the novel conditions of industrial production that seems by its nature to create periodic economic crisis. As we now realize, the startling efficiency of mechanized manufacture created the unprecedented phenomenon of the overproduction of goods such as finished textiles. This oversupply of goods then exceeds demand. Thus the mills must shut down until demand catches up with supply and the unsold goods are purchased. In the early-Victorian period, when mills were shut down or put on a limited schedule, workers were quickly laid-off. Such mass lay-offs were possible since workers were hired without long-term contracts, often on a day-to-day basis. Thus industrial production created the business cycle that in the nineteenth century as today brought about periods of economic depression and mass unemployment.

Such loss or limitation of work was especially destructive since in the early industrial decades there was no government social safety net to protect the workers against fluctuations in employment. There was no unemployment insurance, no welfare provided for families living in their homes, and no government-supported medical care. The only relief came through a system of moving the unemployed or the unemployable such as the aged into workhouses where by law conditions were made so unbearable that there was no temptation to continue on such relief. If one became sick as so many did from brown lung or malnutrition one simply managed until death. If one lost one's job and could not pay for food, one managed or the breadwinner and family simply starved, "clemmed" in Lancashire dialect. The weakening of malnutrition also increased deaths from epidemic diseases.

If in early industrial England the distress of mill-workers and their families was widespread, especially in the 1840s as industrialization began to take off, such hardship had come several decades earlier to those left behind by the rapid pace of mechanical invention. The men who had done weaving in pre-industrial England suffered as much as any other group from the spread of textile machinery. These hand-loom weavers usually worked in their own homes. Jobbers would bring raw material yarn to the individual weavers who would weave the thread into cloth. The jobbers would then return and pay the weavers for the finished fabric. As rapid innovation in machinery pushed the centralization of manufacturing in Lancashire, these artisans could not compete. From skilled craftsmen earning a good living they became outmoded and impoverished. In the Victorian age, these men seemed relics of a lost world. In the nineteenth century, the hand-loom weavers became the emblems of the destructive effects of what we see in our world as technological unemployment.

REGULATING INDUSTRIALISM: WORKER ACTION, GOVERNMENT REFORM, PATERNALISM

In the 1830s and 1840s, the first decades of Victorian mechanization, the dangerous nature of industrial work, the overcrowding and epidemics in the industrial cities, and the poverty generated by the irregularity of industrial employment demanded the attention of the nation. During this time, movements to ameliorate these conditions grew, powered by direct action of the workers, moral sympathy in Parliament, and the effort by certain compassionate mill owners to establish alternative forms of industrial communities. For an industrializing England, these efforts brought about reform without revolution.

Worker Action: Luddism, Chartism, Unionization

The Luddite movement was a direct response by workers against the technological unemployment created by the rise of the machine. In 1811, workers in Nottingham in the north of England, feeling that they had been impoverished by the introduction of new textile machinery, broke into factories to destroy the stocking frames and shearing frames that mechanized the crafts of weaving stockings and finishing cloth. These men claimed to be lead by a mythical King Ludd, named after an actual Ned Ludd who, it was believed, had destroyed stocking frames late in the eighteenth century. This movement of violent machine-breaking by the Luddites spread though the manufacturing districts. In 1812, a group of croppers or cloth finishers put out of work by the new shearing frames attacked a factory and engaged in a battle with armed guards hired by the owner. Luddites also killed the owner of a nearby mill. A contemporary newspaper account describes one such attack in 1812: "A body of men, consisting of from one to two hundred, some of them armed with muskets with fixed bayonets, and others with colliers' picks, who marched into the village in procession, and joined the rioters. At the head of the armed banditti a man of straw was carried, representing the renowned General Ludd whose standard bearer waved a sort of red flag."[8] Such attacks on newly mechanized mills and the homes of the owners continued through the decade.

Alarmed at the threat to textile production, the national government passed the Frame-Breaking Act in 1812 that made the destruction of lace-making machines or stocking frames a crime punishable by death. Although the general feeling of Parliament ran strongly against the workers, some, such as Lord Byron, the Romantic poet, then a member of the House of Lords, spoke against the Frame-Breaking Act. His words show an empathy, rare in the time, with workers suffering from technological unemployment: "Whilst these outrages must be admitted to exist to an alarming extent, it cannot be denied that they have arisen from circumstances of the most unparalleled distress: the perseverance of these miserable men in their proceedings, tends to prove that nothing but absolute want could have driven a large, and once honest and industrious, body of the people, into the commission of excesses so hazardous to themselves, their families, and the community."[9] Still, the government responded with great severity. Troops were sent in to maintain order. There were mass arrests, executions of weavers, and transportation of a number of workers to Australia.

By 1817, the Luddite movement was no longer active, although there were incidents in the 1830s of the breaking of new threshing machines by agricultural laborers. The demise of Luddism can be attributed to changing economic conditions. While impoverishing workers in pre-industrial crafts such as cloth finishing and stocking weaving, mechanization of textile production early in the nineteenth century did create new jobs in the industrial cities that absorbed those whose occupations had disappeared. And yet the movement does live on, at least in language. The term Luddite has come to describe the opposition to technological innovation that continues into our own day.

Along with the violence of Luddism, in the early industrial decades there emerged from the industrial working class protests that called for change through generally nonviolent means. By the 1830s, many factory workers participated in movements calling for unionization, what the nineteenth century termed "combination." Of these attempts by the working class to combine, the most important was the Grand National Consolidated Trades Union founded in 1834 which came to number half a million supporters, mostly of them engaged in poorly paid forms of labor. The aim of such worker organizations was to provide leverage with the owners in bargaining on matters of wages, hours, and working conditions. It should be noted that the idea of a union organized by workers with policy set by workers assumes an inherent opposition of interests between workers and owners. For these early unions, the ultimate weapon for enforcing their demands was the strike, the refusal to work. And in the 1830s, the Grand National Union did engage in a number of strikes, most of which failed.

Even if the unions faced difficulty in recruitment and failure in their use of the strike, the potential power of such worker combinations was greatly feared by the owners and their allies in the government. In these early industrial decades the working class did not have the right to vote and thus had little representation of their interests in Parliament. In 1825, Parliament passed the Combination Act that allowed the trade unions only to meet with owners to bargain over wages and conditions. Any actions outside such limits, such as strikes, made participants liable to prosecution for criminal conspiracy in restraint of trade. In a celebrated case of 1834, the local government sentenced six agricultural laborers of Tolpuddle in Dorset to transportation to Australia for administering what were considered illegal oaths to fellow members of an agricultural laborers group organized to combat a reduction in wages. This highly publicized case of the Tolpuddle Martyrs led to a huge petition campaign from the working class that eventually led to the laborers' exoneration and return.

With the collapse of the Grand National Consolidated Trades Union that had depended on a strategy of strikes and with the success of the action through petition for the Tolpuddle Martyrs there emerged in the 1840s the most significant working-class reformist movement of the time, Chartism. This mass movement focused not on industrial work stoppages, but on presenting to Parliament what was termed the People's Charter, hence the name Chartism. This national petition signed by working people asked the government to pass legislation that would provide entry for working men into the political process. The Charter called only for moderate political reforms allowing the workers the vote and to serve in Parliament, rather than for economic reforms in such matters as wages and hours. The idea was that once allowed political participation, the workers could then promote reformist action to improve working conditions and public health in manufacturing towns. The main demands of the Charter, then, were: votes for all men; abolition of the requirement that Members of Parliament be property owners; secret ballot in elections; and payment for Members of Parliament. In contrast to the Luddite movement and to the strategy of strikes, Chartism manifested a faith in a community of interest rather than conflict between the workers and the government as well as a confidence that the government would accept such moderate reforms for expanding parliamentary democracy. With this trust, the Charter drew 1,200,000 signatures.

The Charter was presented to the House of Commons in 1839; Parliament refused to accept the petition by a vote of 235 to 46. Their faith shattered, the Chartist leaders then called for a general strike, which failed. Demonstrations broke out in Birmingham and other industrial towns. Some Chartists were transported, some killed by the military. A second petition with three million signatures was rejected in 1842; the rejection of the third petition in 1848 brought an end to the movement.

Governmental Reform

Chartism failed, but its failure did not wholly destroy the widespread faith of the British working class that the conditions of industrial labor could be ameliorated by government action. Indeed, under the pressure of working-class movements such as Chartism and motivated by a moral sympathy primarily for vulnerable women and children as well as by fear of working-class revolution, Parliament did move slowly to regulate and reform the conditions of work and of life in the industrial cities. A series of reformist acts were passed beginning in the 1830s and continuing through the Victorian period. It was this continuous moderate reform that spared England from the violent working-class uprisings that occurred on the European continent during the century of industrialization.

In 1832, Parliament held a commission of inquiry on industrial conditions. These hearings in which workers themselves testified brought to public awareness the horrific conditions of child labor and of women's work in the mills and the mines. The testimony provides a vivid impression of how children suffered in the mills:

> What time did you begin to work at a factory?—When I was six years old.
> What was your business in that mill?—I was a little doffer.
> What were your hours of labour in that mill?—From 5 in the morning till 9 at night, when they were thronged [busy with work].
> For how long a time together have you worked that excessive length of time?—For about half a year.
> What were your usual hours when you were not so thronged?—From 6 in the morning till 7 at night.
> What time was allowed for your meals?—Forty minutes at noon.
> Had you any time to get your breakfast or drinking?—No, we got it as we could.
> And when your work was bad, you had hardly any time to eat it at all?—No; we were obliged to leave it or take it home, and when we did not take it, the overlooker took it, and gave it to his pigs.
> Do you consider doffing a laborious employment?—Yes.
> Explain what it is you had to do?—When the frames are full, they have to stop the frames, and take the flyers off, and take the full bobbins off, and carry them to the roller; and then put empty ones on, and set the frame going again.
> Does that keep you constantly on your feet?—Yes, there are so many frames, and they run so quick.
> Your labour is very excessive?—Yes; you have not time for any thing.
> Suppose you flagged a little, or were too late, what would they do?— Strap us [beat with a leather strap].
> Are they in the habit of strapping those who are last in doffing?—Yes.
> Constantly?—Yes.
> Girls as well as boys?—Yes.
> Have you ever been strapped?—Yes.
> Severely?—Yes.[10]

From such testimony of children emerged the Factory Act of 1833. The provisions of the bill, as mild as they seem, illustrate how grim were the conditions of child labor that demanded reform. The Factory Act of 1833 excluded children under nine from working in textile mills and limited children from nine to thirteen to forty-eight hours a week or nine hours in one day with a one-hour lunch break. No person under eighteen could be employed for more than twelve hours a day. Children under thirteen were to attend school no less than two hours a day. Children under eighteen could not work at night. In an important move that limited the power of the factory owners, the act provided for government inspectors to enter the mills to enforce these regulations.

The passage of legislation to regulate and reform industrial production continued through the nineteenth century. The Factory Act of 1844 carried on the improvement of the conditions of employment for women and children in textile mills. Children between nine and thirteen were now limited to working six and one-half hours per day; adult women as well as all persons under eighteen were limited to twelve hours of work each day. The danger of work within the mills was now addressed. Government regulations were introduced requiring the fencing of machinery to prevent industrial accidents. In a continuing move to expand government oversight, all accidents and death caused by machinery were to be reported and investigated.

Early in the 1830s, in the Ten-Hour Movement workers had demanded limitation of the workday to ten hours. Like the early unions and Chartism, the movement failed. But in 1847 the demands of the 1830s were met, in part, by passage of the Ten Hours Act; it limited employment of all women and persons between the ages of thirteen to eighteen to ten hours of work each day during the week and eight hours on Saturday.

As moderate as the Factory Acts were, these Parliamentary actions did assert the principle that in some limited areas government regulation was necessary to moderate the effects of an unchecked industrialization. The lives of the most vulnerable had to be protected. The labor of children was to be limited so that schooling could be provided. The working hours of women were to be controlled. Machinery was to be fenced to prevent maiming on the factory floor. Crucially, the laws were to be enforced by factory inspectors who by gaining access to the mills gained authority over the factory owners. Such government regulation met continued and intense opposition from the mill owners, who adopted a severe position opposing any interference in their operations. Crucially, the Factory Acts broke with the dominant industrial ideology of laissez faire.

Laissez faire, "leave alone" in French, is the theory that government should intervene as little as possible in the economic system so that the economic order can operate according to its own inherent laws. It was this economic theory that allowed both the rapid rise of the machine and its often deadly human consequences. Adam Smith, the celebrant of the theory of the division of labor so central to industrial production, spoke in *The Wealth of Nations* of the "Invisible Hand" or working of natural laws within the free-enterprise system that would promote innovation and economic growth if only government stayed out. In nineteenth-century England, in contrast to the more authoritarian and centralized nations such as France, the absence of government control did free the inventive and commercial energy of inventors and mill owners during the intense period of technological innovation in the late eighteenth and early nineteenth century. And the motive of entrepreneurial gain did spark the intense individual energy of innovators and industrialists.

At the heart of laissez faire is the belief that like the natural world the economic world is governed by built-in laws, and that only harm would result from working against such laws. The Victorian version of this doctrine is well summarized by Thomas Babington Macaulay, a prominent conservative author and politician. Macaulay wrote in 1830: "Our rulers will best promote the improvement of the nation by confining themselves to their own legitimate duties, by leaving capital to find its most lucrative course, commodities their fair price, industry and intelligence their natural reward, idleness and folly their natural punishment, by maintaining peace, by defending property."[11] In one example to justify paying low wages, Victorian industrialists invoked what was called the Iron Law of Wages. Put simply, it was believed that within a free market, the Iron Law would inevitably drive wages to the bare subsistence level, but no lower. Given the competition between individual workers for employment in a free-enterprise system, wages would naturally sink to a low level. Within the free enterprise system, however, wages would by natural economic law never sink below the point needed for subsistence since then the worker would starve and thus be of no use to the employer.

The Iron Law seemed to apply since Victorian industrial ideology saw employment as governed solely by the laws of supply and demand. When there was an oversupply of workers, wages would decline through competition for jobs. When there was a scarcity of workers, wages would rise due to the owner's need for employees. This system assumed that government interference through such acts as enforcing a minimum wage or providing unemployment insurance would only disrupt the working of the economic laws of the market. Furthermore, the owner had no responsibility to the worker beyond the payment of wages when labor was needed, no responsibility to the worker in times of overproduction and economic depression when labor was no longer needed. Following this ideology, early in the nineteenth century the government agreed not to interfere in the working of the economy even in the cyclical downturns that seemed inevitable with mechanized industrialization. Within this economic model, unionization was attractive to the workers as a way of making the struggle between capital and labor over wages, hours, and working conditions more of a competition between equals.

Outside the mill, laissez-faire ideology restrained government action to control the new conditions of industrial life such as unhealthy lodging, air and water pollution, and epidemics. Public testimony as to the long hours that children worked in the mills and the maiming of these children by machinery motivated the Factory Acts. Similarly, the outbreak of contagious disease in the industrial towns and in London also generated a move from strict laissez-faire principles to the realization that local and national governments had a responsibility to modify the harmful effects of industrialization. The exemplar of this government policy shift toward reform was the sanitation movement or what we now call maintaining public health. After epidemics of cholera and typhoid in London, Edwin Chadwick was asked by the national government to report on conditions and causes. His report of 1842, *The Sanitary Conditions of the Labouring Population,* set out the then-radical view that sickness was not the result of fate or personal weakness, but of environmental conditions. Among other reforms, Chadwick urged that localities install systems to provide clean water by pipes and to replace underground pits with sewers to carry off waste. Such suggestions met with strong opposition from property owners and from politicians who resisted government regulation on laissez-faire principles as well as out of economic self-interest. But, as with the Factory Acts, government moved slowly to assert the principle that it must control the effects of the urbanization created by industry. Such legislation gained middle-class support since owners realized that there was an economic benefit, in that the health of industry depended on the health of the workers. Furthermore, it became clear to all that sickness could spread from working-class neighborhoods to middle-class districts. In 1848, Parliament passed a Public Health Act that provided for the formation of a Central Board of Health. This new body had powers to create local boards to oversee street cleaning, refuse collection, water supply, and sewer systems.

But the progress of sanitation in working-class districts was slow. It was not until the 1850s that Manchester had drinking water that came not from the polluted Irk but from a reservoir built outside the city. But even then, the working class had to wait in line at street standpipes to obtain this water. By the 1870s, only a few homes in Manchester had replaced their privies with water closets and these ran directly into the river. In the late nineteenth century, cholera epidemics were still common in Manchester and in other industrial cities.

By the mid-nineteenth century, then, the social management of industrialization in England could be described as a shifting dynamic between laissez-faire entrepreneurship and government regulation. Innovation and industrialization continued to draw energy from the possibilities of individual self-interest. Competition continued: between capitalists for markets, among workers for jobs, and between the unionized workers and the owners. Looking toward gradual reform, local and national governments worked to lessen the danger of factory work, prevent the exploitation of vulnerable groups such as children and women, and to improve public health.

Paternalist Industrial Communities

As England industrialized, some mill owners broke with the dominant laissez-faire model to offer an alternative based on the responsibility of the owner for the general well-being of his employees and their families. Within these communities, the mill was still controlled by the owner rather than by the workers or by the national government. The operating principle was not that of socialism. Rather the owner sought to use his power in a beneficent way, as a father would for his children, thus the term paternalism. It was also assumed that with the benefits of such paternalism, the workers would do their best for the mill by working diligently, not forming unions, and certainly not striking. Such was the vision held by these owners of a community of mutual responsibility rather than the individualistic war of all against all in laissez-faire industrialism.

The nineteenth century saw several examples of such paternalistic communities. The most famous was New Lanark, established by Robert Owen. Looking for a site to put into practice his vision of a humane industrialism, at the beginning of the nineteenth century Owen bought the extensive and highly profitable water-powered cotton mills originally planned by Richard Arkwright at the town of New Lanark in Scotland. Owen's philosophy was that profit-making through cotton-spinning was not incompatible with the benevolent treatment of the workers. To this end he raised wages and improved working conditions within the mills. Before the passage of the Factory Acts, he stopped employing children under ten and reduced the labor of older children to ten hours a day. New Lanark provided clean and spacious housing for the mill-workers and enforced public health within the town. Furthermore, in a time before government-supported compulsory education, he believed strongly in early childhood schooling. The young children of New Lanark went to the nursery schools that Owen built. Older children worked in the factory, but also had to attend his secondary school for part of the day. He also created infant schools, what we would call day care, so that children after the age of one could be well taken care of while their mothers went back to work in the mills.

This community of well-treated, well-housed, healthy workers with high moral standards and little drunkenness who worked in a highly profitable mill provided a sharp contrast to the industrial slums and ragged children of industrial cities such as Manchester. But it must be remembered that the system of paternalism was hierarchical and authoritarian. All rules were made by Robert Owen, the beneficent owner who totally controlled the community. Interestingly, Owen sought to expand his ideal of a paternalist manufacturing community to America. In 1825, he bought the Indiana town of Harmony, renamed it New Harmony, and tried to establish a utopian community there that abolished private property and provided free education for all. Because of internal dissent, the experiment had failed by 1829.

In spite of the success of New Lanark in the formative years of industrialism, the model of the paternalist community did not take root in England. With the movement toward combination that began in the 1830s, workers turned away from Owenite ideas of beneficent authority. Instead, they preferred the working-class unions through which they could bargain on the issues of wages, hours, and working conditions rather than passively accept the authoritarian rule of the owner, no matter how beneficent. As the competitive model of industrial relations became dominant, paternalism lost its force.

Yet one example of a successful paternalist community can be cited in the late Victorian period. In 1888, William Lever, the head of Lever Brothers, a successful soap manufacturing corporation, created a model village and manufacturing plant at a site outside Liverpool. He named the community Port Sunlight, in honor of his best-selling soap brand, Sunlight soap. The new factory was clean and well-ventilated, with regular medical inspections for the workers. He donated part of the company profits to build and maintain worker's housing modeled on the late-Victorian ideal of a garden suburb of individual homes surrounded by public parks. He also invested in the cultural life of the workers' community, building a concert hall and a large art museum that he felt would contribute to the aesthetic development of his employees. His ideal, like those of early-Victorian paternalists, was to revive what he imagined as the sense of community that existed before the laissez-faire individualism of industrialism. He wanted, in his own words, to "Christianise business relations and get back to that close family brotherhood that existed in the good old days of hand labour."[12]

Yet, as benevolent as Lever's intentions were, his practice was utilitarian as well as authoritarian. Like Robert Owen, Lever wanted the workers to be happy and healthy so that they would be productive. He believed in profit-sharing, but rather than sharing profits directly with the workers, he invested the proceeds from his corporation in Port Sunlight in ways that he himself saw as benefiting his employees. He had little faith in the ability of the workers to keep from drink and bodily pleasure. He said to them about raising wages, "It would not do you much good if you send it down your throats in the form of bottles of whisky, bags of sweets, or fat geese at Christmas. On the other hand, if you leave the money with me, I shall use it to provide for you everything that makes life pleasant—nice houses, comfortable homes, and healthy recreation."[13] Although no longer reserved for Lever workers, Port Sunlight remains today a delightful community of lovely small houses with a vibrant community life.

THE INDUSTRIAL NOVEL

The social effects of laissez-faire industrialism also generated calls for reform from prominent novelists of the time. In the 1840s, there came into being the industrial novel, a new form of fiction seeking to show ways to ameliorate the conditions of factory labor and to diminish class conflict.

Chief among these novels are: Frances Trollope's *Michael Armstrong, the Factory Boy* (1840); Charlotte Elizabeth Tonna's *Helen Fleetwood* (1839–41); Benjamin Disraeli's *Sybil: or, The Two Nations* (1845) and *Coningsby: or, The New Generation* (1844); Elizabeth Gaskell's *Mary Barton* (1848), *Ruth* (1853), and *North and South* (1855); Charlotte Bronte's *Shirley* (1849); and Charles Dickens's *Hard Times* (1854).

The industrial novel focuses on the social questions that appeared most important to the first Victorian generation to confront industrialization. Were the mills to be governed by the paternalism demonstrated in Owens's New Lanark or by laissez-faire principles that looked to the invisible yet ultimately benevolent hand of the free market? Should owners and workers act according to calculated self-interest or seek the well-being of the community as a whole? Should relations between workers and masters be based on cash payment or on personal friendship?

In addressing these questions, the novels are structured according to the pervasive early-Victorian sense that industrial England had become divided into what Disraeli in *Sybil* called the "two nations." For Disraeli the two nations are the rich and the poor, the owners and the workers. For Gaskell in *North and South*, the split is between the industrial North represented by Manchester, with its faith in economic self-interest, and the pre-industrial agricultural society of the South of England, based on personal relations and mutual responsibility.

The social aim of these industrial novels is to avert war between the two nations. Underlying these novels is the middle-class fear of a working-class revolution that is represented in the narratives by the workers' strike, usually depicted as violent physical struggle. Writing for middle-class readers, these middle-class authors advocate reform rather than radical change in the economic system. Like the paternalist reformers, none advocate sharing ownership with the workers. Generally, the novels look to ways of ameliorating life outside the factory rather than a transformation in the modes of mechanized production within the mills. In *Hard Times*, for example, Dickens sees the imagination, represented by a traveling circus, as providing rest and recreation for the workers from the inevitable strain of factory labor.

In these novels, reconciliation of industrial conflict is achieved through the personal rather than the political. Typically, a figure from the world of the owners journeys into the land of the working class to form a bond of sympathy with individuals, as do Margaret Hale in *North and South* and Louisa Gradgrind in *Hard Times*. These novels often end with a marriage between representatives of the opposing worlds that signifies the possibility of modifying the industrial system without conflict. In *North and South*, Margaret Hale, the daughter of a minister from the South of England, moves with her father to Manchester (the North) where she eventually marries John Thornton, the owner of the mill that has been subject to a violent strike. The novel suggests that this marriage will allow Margaret to understand the economic responsibilities of the master of the mill, while using a fortuitous inheritance to help him create a paternalistic community. Uniting the interests of the owner with the interests of his workers will prevent future strikes.

Industrial novels, then, tend to resolve class conflict through marriage. It could be argued that such endings with their emphasis on the personal divert attention from social reforms such as transferring political and economic power from the owners to the workers. These narratives that turn from the male sphere of economic calculation and labor unrest to the female sphere of emotional connections call attention to the fact that many exemplary industrial novels were written by women—Gaskell, Charlotte Bronte, Frances Trollope, Tonna. Furthermore, these novels often have female protagonists, as seen in such titles as *Mary Barton, Sybil, Shirley, Ruth,* and *Helen Fleetwood.* These works, then, imagine a way for women to enter the male public sphere of industrial relations so as to moderate the harshness of laissez-faire self-interest.

For all its energy, the industrial novel had a short life. By the 1860s, industrial novels were no longer written. Several reasons can be given. For one, the conditions of industrial labor as well as the landscape and the life of the industrial city had become so familiar that middle-class readers no longer sought fictional tours of the factory world. Although struggles about control of the industrial system continued, as did strikes, these conflicts had lost the aura of novelty that draws readers to novels. The fundamental arguments about the need for class reconciliation lost their edge as the nation accepted the idea of continued conflict between capital and labor. Finally, the possibility of violent working-class revolution in England diminished with the gradual providing of workers with the vote, the growth of unions, expanding governmental regulation of industrial life, the increasing general wealth provided by mechanized manufacturing, and the betterment of public health through sanitary reform.

6

Late Victorian Invention

In the first half of the nineteenth century, England led the world in technological innovation. As the premier industrial nation, England developed the
factory system, synergistically fusing the railway, the steam engine, and
automatic machinery to generate an abundance of mass-produced consumer
goods. Globally, the steamship, the undersea telegraph cable, and the mechanization of warfare enabled this small island to rule a vast empire that provided raw materials for English factories and markets for England's
manufactured goods. In 1851, England celebrated its primacy in the age of
the machine by staging the Great Exhibition of the Works of Industry of
All Nations in the Crystal Palace.

But by the later decades of the Victorian age, the term all nations had a
new significance, for technological innovation was no longer the sole prov-
ince of the experimenters and industrialists of England. Instead, the rise of
machine technology had accelerated among a group of newly industrializing
nations including France, Germany, and the United States. These nations
were approaching the takeoff point toward fully mechanized production and
were devoted to challenging England's technological supremacy. This global
spread of invention took the form not of an arms race, but an invention
race. Nineteenth-century countries realized that national prosperity and
power were linked not merely to the strength of armies, but, as today, to
maintaining a technological edge that generated wealth and well-being. In
these industrializing nations, and even in England, government increasingly
supported the work of technologists and scientists.

In part, this globalization of invention, so similar to the international-
ism of science and technological development in our own time, emerged

inevitably from the very dynamic of invention that England had pioneered. Paradoxically, the technologies created by England early in the Victorian period facilitated the spread of invention beyond England. Scientific and technical knowledge, as we now realize, cannot be kept secret. The development of virtually instantaneous communications between nations through the telegraph lines that ran under the sea enabled the swift spread of technological information. With rapid transportation between countries by railway within the European continent and steamships across the oceans, inventors and technologists were no longer confined to their home countries, but able to travel to nations that would support their work. For example, the Italian, Guglielmo Marconi, relocated to England where he could find support from the British government for his work in wireless telegraphy. In the essential dynamic of technological change, innovations accrued in a gradual process of improvement. But from the 1860s, this process became global rather than national. Innovation occurred, often simultaneously, throughout the industrialized world. These innovations were rapidly communicated and swiftly incorporated into new technological systems.

As the process of invention continued in the global arena, technology in all nations moved beyond the utilization of steam that England had pioneered and that was the source of its early industrial supremacy. The age of steam gave way to the age of electricity. The energy of invention gave rise to a new technological world in which electricity provided power to factories, illuminated homes, and powered communication by telephone and eventually by wireless. Another source of power, the self-contained internal-combustion engine running on forms of oil, enabled the building of the automobile and the fulfillment of the ancient dream of manned flight.

The story of Victorian technology in the latter nineteenth century, then, is no longer the narrative of England as supreme in the invention and application of the machine. In these decades, England continued its dynamic of technological innovation, but as one among many competing industrial powers. In some fields, such as the theory and the practical machinery of electrical generation, English scientists and inventors remained pioneers. In other areas, such as the development of the wireless telegraphy that became radio, England hosted and supported a foreign inventor so as to become the leading site of such innovation. In still other fields, such as electric lighting, English inventors made significant contributions that occurred simultaneously with work in America. In the invention of the telephone, the Scottish-born Alexander Graham Bell did his major work in America. The leading work in the development of the internal-combustion engine took place in Germany, yet credit must be assigned to a British engineer for inventing what came to be known as the diesel engine. The first successful heavier-than-air flight took place in the United States, yet drew upon the writings of an early-Victorian British experimenter.

As Queen Victoria's reign ended in 1901, the technologies that came to dominate the twentieth century and still flourish in the twenty-first century were firmly in place in England—clean electric power, electric lighting, voice communication over the telephone, wireless communication, the phonograph, the internal-combustion engine, and the automobile.

THE AGE OF ELECTRICITY

In the first half of the nineteenth century, Victorian mills and factories were powered by the coal-burning steam engine whose soot hung over the industrial cities. Victorian cities were bound together by the steam locomotive. But by the late nineteenth century, many factories were run not by the energy of expanding steam, but by the power of electricity. Electricity was generated at central power stations and electric current was transmitted by wire to factories and homes. Small electric motors powered machines without the necessity of transmission wires. Electricity replaced steam in underground urban railways. Electric lighting transformed homes and public spaces. The communications revolution of the electric telegraph was continued in the telephone and in wireless communication powered by electric current. Victorians could now listen to mass-produced sound recordings on what the age called the gramophone. If the earlier Victorian period could be called the age of steam, the later Victorian decades could be termed the age of electricity.

Generation of Electricity

The fundamental methods and the workable machinery for generating electricity were first developed in England by the early-Victorian inventor, Michael Faraday. Before Faraday's work in the 1830s, the only means of producing an electrical current was the chemical battery invented in eighteenth-century Italy by Alessandro Volta, after whom our metric of electrical power, the volt, is named. The forerunner of the batteries that power our cell phones and iPods, the voltaic pile, as these early batteries were called, consists of a stack of different metals, such as silver or zinc. The discs are separated from each other by a piece of cloth or cardboard that had been soaked in salt water. The pile generates a small current in a wire attached at each end. For Volta, and other eighteenth-century Italian experimenters such as Luigi Galvani, these proto-batteries functioned primarily for spectacle and philosophical experiment. In one series of experiments, current from a voltaic pile was run through the leg of a frog, causing the leg to twitch. The leg moved through what we now know as the electrical activation of muscles, but for eighteenth- and early nineteenth century thinkers the leg movement seemed to demonstrate that electricity was the long-sought vital life force powering the body. Mary Shelley, author of *Frankenstein*, writes she conceived the book in 1816 with the notion that "a corpse would be re-animated; galvanism [the contemporary word for electrical current] had given token of such things: perhaps the component parts of a creature might be manufactured, brought together, and endued with vital warmth."[1] This notion of electricity as the battery of life continues in the convention of the giant electrical machines that bring life to Victor Frankenstein's assembly of dead body parts in innumerable retellings of Shelley's early nineteenth-century tale.

Occupied with finding the principle of life, these eighteenth-century experimenters had no thought of the industrial and commercial application of electricity. In any case, the voltaic piles were too cumbersome and costly for general use. It remained to Faraday, a practical Englishman living in a nation and in an era that prized practical inventions, to seek a technology that would apply electricity to productive purposes. Like James Watt, the contemporaneous Scots giant of invention, Faraday fit securely into the British mode of inventor-as-tinkerer. A self-taught experimenter like Watt, Faraday had no training in theory. Rather, as throughout the early nineteenth century, invention preceded theory. Just as the Watt steam engine generated the need for a science of thermodynamics describing the relation of heat to other forms of energy, so Faraday's innovations brought forth theories of electromagnetism describing the relation of electricity and magnetism.

Faraday's work emerges from his study of Volta's chemical batteries and especially from the early nineteenth-century discovery of the interconnection of electricity and magnetism. In 1820, the Danish experimenter Hans Christian Ørsted noted an unexpected phenomenon in his workshop. A compass needle was deflected from magnetic north when the electric current from the battery he was using was switched on. This movement convinced him that the electric current running through the wire produced a magnetic field and that it was this magnetic force radiating from the wire that affected the compass needle.

Familiar with Ørsted's work, in the early 1830s, Faraday reasoned that if electricity could generate magnetism, then magnetism could generate electricity. It was this fundamental recognition of the interconnection of electricity and magnetism, now termed the phenomenon of electromagnetism, that is central to the age of electricity. And it is Faraday's very British occupation with the practical and the useful that led him to apply this insight to the construction of workable devices setting out the basic technology that still underlies the generation of electricity.

In crucial experiment in 1831, Faraday developed the fundamental mechanical means of generating electricity by building what he termed an induction ring. He wrapped two separate coils of wire around an iron ring. He found that by running a current from a battery through one coil, the electromagnetic field created in that coil generated an electric current in the other coil. This generation of electricity in a wire by means of the electromagnetic effect of a current in another wire has continued to be called induction and is one of the basic principles of generating electricity.

In another critical experiment of 1831, Faraday constructed what was the first electromagnetic electrical generator, later called the Faraday disc. For this device, he connected two wires to a copper disc, and then rotated that disc through the poles of a horseshoe magnet. A current was generated in the wires connected to the disc. This experiment demonstrated that moving a wire through a magnetic field could generate electricity in the wire. He also found that moving a magnet through a loop of wire caused an electric current to flow in one direction in the wire. In each method, the magnetic field produced electrical current. From these small-scale experiments emerged the large-scale generation of electric power and the electrification of industrial nations.

From Faraday's disc came the dynamo, the machine employed later in the century for generating high voltages on a large scale that could then be transmitted over long distances by wire. Faraday found that higher, more useful voltages could be produced by winding multiple turns of wire into a coil. This innovation is at the heart of the dynamo. In the structure of the large-scale dynamo this winding of wire coils rotates through a powerful magnetic field. This magnetic field is generated either through stationary magnets or electromagnets, devices in which magnetism is created by the flow of electric current. As in Faraday's disc, direct electric current is created in the wires, but given the industrial scale of such machinery, the current is of a high voltage suitable for transmission. The first central stations housing dynamos that generated electricity for transmission by wire were built in England the early 1880s.

For the dynamo to work, an external force is needed to rotate the wire coils through the magnetic field. By the latter part of the century, this motion was supplied through the use of a turbine. A turbine is a simple machine in which the flow of a fluid over blades imparts a rotary motion. The water wheel that powered the first textile mills was a primitive form of turbine. In the late-Victorian period, the mechanics of the turbine were applied to generate electrical power. Here, the winding of coils was attached to a rotor assembly, a shaft with blades attached. Ironically, the methods of the age of steam were utilized to generate this new form of energy as the blades of the turbine were turned by expanding steam generated by the burning of coal, although oil was also used by the end of the century. In 1884, Charles Parsons patented in England a particularly efficient steam turbine in which, as in the early-Victorian factory, the water is heated in a separate power plant. In the Parsons model, the expansive steam ran directly over the blades of the turbine converting the energy of steam to electricity.

Through an elaborate system of differing-size wheels on the shaft of the turbine, the machine could operate efficiently without creating rotation speed beyond the capacity of the materials. Hydroelectric generating plants were also built in which the flow of falling water moved the blades of the turbine.

With the development of large-scale electrical generating devices housed in central stations came the problem of transmitting electric power. The problem of transmission is crucial since generating stations are often located near the sources that power the plants, such as falling water or coal, and thus are far from urban areas. The nineteenth century at first employed direct current, current that flowed in one direction as in the current produced by batteries. This current was sent in overhead lines at the voltage applicable to machinery or urban lighting systems. By the last decades of the century, however, it was found that alternating current or AC, current that reverses direction periodically, usually many times per second, can be sent from the generating plant at high voltage and transmitted over long distances with little loss of energy. In the late nineteenth century, this AC high-voltage current was sent through overhead wires attached to poles. These lines were eventually linked in a grid of interlinked power lines so that current could flow from different generating plants into a single system, making power available even if one plant failed or went off-line. In urban areas the multiplication of these lines on poles running along streets became unsightly, dangerous, and inefficient, so that electrical transmission wires were gradually placed underground.

Transmission lines carry a single form of high-voltage electricity generated at the central station. In the grid, lines run to substations where the current is stepped down or reduced by transformers to suit local needs. The transformer is basically an adaptation of Faraday's induction coil in which the electric current in coiled wire on one side of an iron ring generates a magnetic field that induces or generates current in wire coiled on the opposite side of the ring. In transformers in the power transmission system, the high-voltage AC comes in to one set of coils. The opposite side of the ring contains fewer coils so that a lower voltage can be created suitable for lighting, machinery operation, or other needs.

By the end of the Victorian period, the work of Faraday had been magnified into an electrical system for the generation of electricity through turbines powered by water or coal in central power plants. This current could be efficiently transmitted at high voltage through the nation and adapted to varied needs through substations and transformers. This electrical system provided a source of clean power that gradually replaced the stationary steam engine as the energy source for industry. With the momentum of technological innovation, inexpensive electricity engendered new inventions that in many ways transformed daily life of later Victorian England.

Transportation by Electricity

As in our own automobile age, industrialization brought urban traffic congestion. With the growth of London as the center of an industrial nation, the density of traffic on the streets became intolerable. Horse-drawn carriages and omnibuses carried passengers through the city. Horse-drawn freight wagons carried raw materials to workshops and finished goods to the stores. Pedestrians had to navigate between these vehicles on streets muddied with horse droppings. At mid-century, London engaged the traffic problem by building a rapid transit system in tunnels beneath the streets. By the end of the century, the underground trains were powered by electricity.

The underground transit system, the London Underground as the present extensive subway system of the London metropolitan area is called, was originally conceived as an extension of the aboveground railway network that ran into the urban center. In 1863, the Metropolitan Railway, the world's first urban underground passenger-carrying railway, was completed to run from the busy Paddington Street railway terminus to the City, the financial district of London. Its purpose was to allow passengers to travel from the suburbs to their work in commercial offices without encountering street congestion. The Metropolitan Railway used steam locomotives, as did the other underground lines built at mid-century. Once the underground railway was proved successful, the network rapidly expanded. As in the railway construction boom of the 1840s, private companies looking to make a profit built new lines connecting other areas of London. The separate private lines were eventually merged into the unified network of the London Underground.

The underground system expanded to areas on the south side of the Thames River by incorporating in 1869 a tunnel under the Thames constructed between 1825 and 1843 for pedestrians. The tunnel had been built by Sir Marc Isambard Brunel and his son Isambard Kingdom Brunel, the engineers of the great railway bridges so crucial to the early expansion of the aboveground railway. For this unprecedented project, the first tunnel to be constructed underneath a navigable river, the Brunels, father and son, invented a new method for excavation below water. The Brunel tunneling shield, to quote the contemporary *Illustrated London News*, consisted of

> 36 chambers of cells, each for one workman, and open to the rear, but closed in the front with moveable boards. The front was placed against the earth to be removed, and the workman, having removed one board, excavated the earth behind it to the depth directed, and placed the board against the new surface exposed. The board was then in advance of the cell, and was kept in its place by props; and having thus proceeded with all the boards, each cell was advanced by two screws,

one at its head and the other at its foot, which, resting against the finished brickwork and turned, impelled it forward into the vacant space. The other set of divisions then advanced.[2]

As the miners slowly lengthened the tunnel, bricklayers quickly built the supporting frame of the tunnel as the top, sides and bottom.

The tunnels, or tubes as they were called, that ran underneath the city itself were at first constructed with the cut-and-cover method that had built the tunnels that brought the trains through the neighborhoods of London and into the great railway stations. In this technique, a large trench was excavated on the city street, tracks were laid, then the hole in the city covered over. Cut and cover created enormous disruptions in the city, destroying entire neighborhoods, as Charles Dickens describes in his account of the coming of the railway to London in *Dombey and Son*.[3] (See chapter 1, The Factory System.) By the end of the century, methods of underground drilling using variations of the Brunel tunneling shield were adopted and used with increasing safety as workers could reinforce the walls of the tunnel as the shield moved forward. This tunneling method avoided surface disruption and also allowed the digging of new deep-level lines of the Underground.

The first railway trains to enter London in tunnels and the first underground trains that ran beneath the city emitted noxious, dense, coal smoke from the steam locomotives within a confined space. This coal smoke created serious problems of health and visibility. To clear the smoke, the railways and the Underground built ventilation openings intruding into London streets at regular intervals. In the later Victorian decades, with the extension of the Underground through deep-level lines tunneled far below ground, the system of ventilation shafts reaching to the surface to vent the emissions of steam locomotives was no longer feasible. The answer to this ventilation problem was the electrification of the lines. In 1890, the City & South London line, the first electrically powered underground line, went into operation using electric locomotives that were specially designed to fit the cramped spaces of the tubes. The new electric locomotives were smokeless. These first small electric locomotives were powerful enough to pull a subway train of three cars. These had electric motors installed that drew power from overhead lines and that directly drove the drive trains and thus the wheels of the train. This process of electrification continued until the Underground became fully electrified in the early twentieth century.

Electric Light

By the end of the nineteenth century, the creation of a transmission grid that could deliver electric power efficiently at a range of voltages made possible the application of electricity to the lighting of public spaces, factory floors, and home interiors. The primary technological problem was to create practicable devices that could transform this available electric power into light. As typical in the late nineteenth century, the process of innovation was international, with inventors in England and in America simultaneously reaching solutions. By the end of the century, electric light had replaced coal-gas lighting just as coal-gas lighting had earlier replaced candles and whale-oil lamps.

The nineteenth century saw a competition between two basic methods of turning electricity into light: the arc light and the incandescent light. The invention of the arc light can be attributed to the British technologist Sir Humphrey Davy. Davy is a central figure in the industrial revolution for his invention in 1815 of a safety lamp for coal miners working in mines containing the explosive gas firedamp, the name by which methane was known. This device allowed deep coal deposits to be mined, thus increasing production of the coal so necessary for an industrializing England. Years earlier, in 1801, Davy invented the basic form of the arc lamp. The arc lamp consists of two rods of carbon placed close together. When a current runs through both, the electricity jumps or arcs between the ends of the rods creating a very bright light in the space between. In his first arc lamp, Davy used charcoal sticks and a battery system with 2,000 cells to create an arc of electricity across a four-inch gap. By 1846, a practical form of the arc light was developed in England in which electrodes of carbon were gradually pushed together as they burned to create a constant brightness. The intense quality of light generated by the arc light was useful in certain applications, such as the lighting of the stage in theatrical performances. When suitable electric generators became available in the late 1870s, the widespread practical use of arc lamps became possible. The carbon arc street light was developed for public spaces where brightness was a value. But the need for high voltage, the danger of the open spark, and the high intensity of the arc light made this device impractical for small spaces, such as domestic interiors and small workplaces.

Given the limitations of the electric arc lamp, the late nineteenth century search for a widely applicable form of electric lighting turned to the incandescent solution. The near-universal extension of lighting by electricity came with the invention of what we now call the light bulb or what is more accurately termed the electric incandescent lamp. In this device a filament is placed within the enclosed vacuum of the glass bulb. As the filament is heated to a high temperature by an electrical current, the resistance of the filament causes it to glow and thus emit light.

Although credit for the electric incandescent lamp is often assigned to the American Thomas Alva Edison, the electric filament lamp was actually developed earlier in England. In 1850, the Englishman Joseph W. Swan had begun work on a light bulb using carbonized paper filaments. In 1860, he had built a working bulb and obtained a patent in England for a bulb with a carbon filament within a partial vacuum. In 1875, Swan developed a bulb with a better vacuum and a carbonized thread as a filament. With little oxygen in the bulb, the filament could glow almost white-hot without catching fire. Swan received a British patent for his device in 1878, about a year before Thomas Edison received his American patent.

At a lecture in Newcastle, England, in 1879, Swan demonstrated his incandescent electric light bulb in front of an eminent audience of the Newcastle Literary and Philosophical Society. He had 70 gas jets turned down and their light immediately replaced by 20 electric bulbs. That year, he began installing light bulbs in homes and landmarks in England. In 1881, he started his own company, the Swan Electric Light Company, and a factory for the mass production of light bulbs. When Edison attempted to market his own filament bulb, also using a carbonized thread, in England in the early 1880s, a complex patent fight erupted with Swan. Finally, the Edison and Swan United Electric Company was created in England to market bulbs jointly. The use of the incandescent bulb spread, eventually replacing gas lighting in the home and in public spaces.

Like any other invention, the incandescent light bulb can only function within a technological system. And from the innovations of Swan in late nineteenth century England a system of electric lighting came into being. This lighting system, so familiar to us, depends upon central power plants generating current with their turbines; transmission lines moving the electricity from the power station to the urban areas; transformers in substations to step down the voltage with Faraday induction coils to the level appropriate to homes and businesses; as well as fixtures to hold the electric bulbs and hardware such as sockets and light switches.

Telephone

The primary inventor of the telephone, Alexander Graham Bell, exemplifies the global movement of inventors and invention in the late nineteenth century. Bell was born in Scotland and educated at the University of Edinburgh. In 1870, at age 33, Bell moved with his family to Canada and shortly thereafter to the United States. Only in 1882 did Bell become a naturalized American citizen.

The invention of the telephone continues the application of electricity to communication in the Victorian age. As the transmission of sound, especially speech, by wire the telephone expands the function of the electric telegraph. Indeed, the Victorians described the telephone as the speaking telegraph. The electric telegraph, so crucial to regulating the railway system and to transmitting news and messages in the early-Victorian age, is powered by electricity that is generated by batteries. These batteries maintain a constant current running through the wires. The use of a key at one end of the electric circuit turns the current on and off to produce in the receiver at the other end a series of long and short clicks read as the dots and dashes of Morse code. Yet human speech does not consist of dots and dashes, but is a smooth yet varying flow of sound. The primary challenge for nineteenth-century inventors seeking to transmit speech was to create variations in a continuous electrical current moving through the wire so that a mechanical device at the other end of the circuit could reshape the current back into tones reproducing the original sounds of speech.

Having worked on the possibility of a mechanical ear for deaf people, Bell modeled his telephone on the human ear, theorizing that variations of electrical current replicating the changes in waves transmitting sound through air could replicate speech by acting on a mechanical membrane substituting for the human eardrum at the other end of the circuit. His 1876 patent filed with the United States Patent Office described "the method of, and apparatus for, transmitting vocal or other sounds telegraphically ... by causing electrical undulations, similar in form to the vibrations of the air accompanying the said vocal or other sound." In Bell's first transmitting device, sound waves moved a diaphragm over the top of a jar filled with acid electrified by a battery. The movement of the diaphragm raised and lowered a steel rod inserted in the acid, thereby raising and lowering the acidity and thus the resistance to electricity of the liquid. The varying current passed through a wire connected to a membrane at the other end that vibrated in tune with the current and thereby transformed the current into a replica of speech. Using this liquid transmitter, in 1876, Bell made the first telephone call over 100 feet of wire. Since this was the first telephone line, there was no busy signal and he was able to summon his assistant immediately from the basement with the celebrated words, "Mr. Watson, come here, I want to see you." This was the first instance of the transmission of speech over a wire using the power of electricity.

The cumbersome acid liquid method of varying the current was soon replaced by other methods of transmitting sound. In 1878, Thomas Alva Edison developed the carbon transmitter that in its basic form remained in use into the twentieth century. In this device, loose carbon grains are set between two metal plates and a direct current is passed between the plates through the carbon grains. The plate facing the speaker serves as a diaphragm so that the variations in sound cause this plate to move back and forth compressing and loosening the grains. As the density of the carbon grains changes, the resistance between the two iron plates varies and thereby varies the current. This continuously varying current passes as electrical impulses through wire to a similar carbon microphone at the other end that replicates speech by moving another sound plate.

In this time of globalization, inventions did not long remain the possession of one nation. The telephone came with the speed of electric current to
England. In 1877, one year after Bell filed his United States patent for the
telephone, Bell himself demonstrated the telephone to Queen Victoria, making several calls from the queen's residence on the Isle of Wight to London.
In that year, Sir William Thompson (later Lord Kelvin) exhibited Bell's telephone to the British Association for the Advancement of Science. Quite correctly seeing the telephone as fundamentally an outgrowth of the telegraph,
he described it as "the greatest by far of all the marvels of the electric telegraph."[4] Indeed, the successor to the electric telegraph used the infrastruc-
ture of the older system. Commercial long-distance calling in England began
in 1878 with a call between London and Norwich, a distance of 115 miles,
using already existing telegraph wire.

Like all inventions, the telephone generated further innovations that came
to make up the telephone system. The first telephones were connected by a
single wire from one location to another, much like the line connecting the
first floor and the basement workshop of Bell's home. In the early days of
the telephone, a person would set up a direct line between, for example, his
office in central London and his office at the London docks. The need to construct several lines if one wished to speak to several locations and the cost of
maintaining them led to a method of establishing telephone exchanges similar to the very successful telegraph exchanges. In the telegraph exchange
model, one would go to an exchange office to send a telegram that would go
to another exchange office. Once received in that exchange, the telegram
could be delivered by hand or picked up by the recipient.

The system of the early-Victorian telegraph exchange was reworked for
the telephone. In 1879, the Telephone Company using Bell's patents opened
Britain's first public telephone exchanges with a total of 200 subscribers.
Public phone exchanges soon opened in the manufacturing cities of the English Midlands. In this system, individual homes or business were connected
to switchboards located in a central telephone office building. The caller
would pick up the phone, a light would come on at the switchboard, and the
caller would request a number to be called. The operator would then connect
the caller to this number. The first automatic central exchange, an exchange
that employed machinery rather than human operators to connect callers,
was set up in England in 1883.

In 1879, English law ruled that the telephone was a form of the telegraph
and thus came under the jurisdiction of laws governing the telegraph that
were applied by the British Post Office. Private companies had to apply to the
Post Office for permits to operate. Working together, private companies and
the Post Office generated a number of innovations. As the British system
expanded, call offices were established for the use of persons who could not
afford to own a phone. Kiosks, the British term for phone booths, were also

built. Telephone directories were published. Electrical power for the phone was supplied by a battery attached to each phone. In some cases, the subscriber had to crank the phone to generate the needed power. Eventually, the low current necessary was supplied through the phone network along the same telephone lines that carried the varying electrical impulses replicating sound.

As with the telegraph and the railway, the web of communication quickly broadened. The phone network spread rapidly. Long-distance lines, called trunk lines, extended across the nation, much like the trunk lines or main lines built during the early-Victorian railway boom. Telephone communication was opened between London and the Midlands when in 1890 a trunk circuit linking London to Birmingham was brought into service by the National Telephone Company. In 1891, the first submarine telephone cable was laid across the English Channel, enabling telephone conversations between London and Paris. By the end of the Victorian period, then, the telephone system in its recognizable modern form had become firmly and widely established in England.

Wireless Telegraphy: The Forerunner of Radio

As it became clear that an electrical current could carry sound over wire, it became imaginable that electrical current could transmit sound without the need of a wire. Working from innovations going back to Faraday and from the theorizing of electromagnetism by a British physicist at mid-century, inventors in England developed what the late nineteenth century called wireless telegraphy, the communication of the familiar dots and dashes of the electric telegraph, but without wires. From this wireless telegraphy came the communication of sound and speech that in the twentieth century was eventually called radio.

Faraday's experiments of the 1830s had showed that magnetism could generate electric current in a wire and that, conversely, an electric current could induce magnetism in a wire. Faraday had moved in characteristic British fashion from the experimental to the theoretical. In the early 1860s, innovation moved from the theoretical to the practical. The British physicist James Clerk Maxwell developed a fundamental and powerful set of equations describing the phenomenon of electromagnetic waves. These equations showed how electricity, magnetism, and even light could be described within a unified field called the electromagnetic field. Maxwell's equations were the most significant contribution to physics in Victorian England. In 1931, on the centennial of Maxwell's birthday, Albert Einstein described Maxwell's work as the "most profound and the most fruitful that physics has experienced since the time of Newton."[5]

After the publication of Maxwell's equations in the 1860s, experimenters around the world sought practical proof of the existence of the electromagnetic waves predicted by Maxwell. Such tangible verification was finally provided by the German physicist Heinrich Hertz, today commemorated in the "hertz" used as the unit for measuring frequency. Hertz sought to demonstrate that high-energy electromagnetic radiation or radio waves could travel through space without wires. To demonstrate, he built an apparatus called the spark-gap transmitter, in which high voltage ran at regular intervals through a wire with brass knobs at each end. Running the current through the wires created high-voltage sparks that arced the gaps between the brass knobs. To receive the waves generated by the arcing sparks, he built a receiver of similarly separated brass knobs again placed on the ends of a wire. This receiver was placed several yards from the transmitter. When the transmitter sparked over the gap, a current was induced in the receiver, creating sparks between the brass knobs. This foundational experiment proved both the existence of electromagnetic waves and that these waves could be transmitted over distance.

The existence of these electromagnetic waves traveling through space without wires and with the speed of light offered the possibility of virtually instantaneous communication. With the swift international transmission of knowledge in the later nineteenth century, inventors around the globe worked to find practical applications of electromagnetic principles to communication without wires. The primary credit for a workable system of inventions to make wireless practical goes to Guglielmo Marconi, whose work in late-Victorian England with British government support brought into being wireless and eventually radio.

Born into a wealthy Italian family, as a young man Marconi experimented with systems of wireless following the work of Hertz. In 1895, he succeeded in sending wireless signals over a distance of one and a half miles. But finding little interest in such technology in Italy, at the age of 21, Marconi moved to England. There, he attracted the attention of the chief engineer of the British Post Office, the government agency with responsibility for telegraph and telephone service in England. As the term wireless telegraphy suggests, Marconi and the British Post Office, in looking for a wireless system, were seeking a telegraph system without the need of telegraph wires. In 1896, Marconi was granted a British patent for an invention by which "electrical actions or manifestations are transmitted through the air, earth or water by means of electric oscillations of high frequency." This was the world's first patent for a system of wireless telegraphy. Marconi was a practical man and in 1897 he established, albeit after innumerable patent fights, the Marconi Wireless Telegraph & Signal Company.

The late-Victorian wireless patented by Marconi differed from the telephone and from radio as we now know it in not aiming to transmit speech through variations in an electric current. Instead, the goal was a transmitter that sent electrical charges through the air so that by shifting the current on and off in a receiver, these charges would translate into the dots and dashes of Morse code. Hertz's use of an oscillator in his experimental transmitter to regulate the electrical charges passing through the wire had pointed ahead to ways of controlling signals in an on-and-off fashion similar to that of the electric telegraph. For his practical wireless system, Marconi drew upon the work of Hertz, upon the methods of the existing electric telegraph system, and upon his own innovations. For his transmitter, Marconi used the spark-gap device developed by Hertz, but with the innovation of setting the gap between an antenna and the ground. Using the increased electrical power available by late century, he was able to harness pulsing bursts of electricity to create a spark of up to ten thousand volts. The spark could send out a burst of electromagnetic radiation that traveled a long distance without wires. Since the wireless telegraph was just that, an electric telegraph without wires, Marconi added to the system a telegraph key for the transmitter to offer dots and dashes in the manner of the early-Victorian telegraph. In the receiver, the powerful spark from the transmitter would generate a current that would arc the gap, thereby closing the circuit as does a switch. As the burst stopped, the circuit would reopen.

The receiver within the wireless system depends upon an innovation called the coherer, a glass tube filled with filings of silver and nickel. An improved iron-and-mercury variation was used by Marconi for the first transatlantic radio message. In a manner analogous to the varying of electric current in the carbon grain telephone transmitter, the filings in the coherer become more tightly packed and offer less resistance to electricity when subject to the high energy transmitted by the improved Marconi spark-gap transmitter. With this lowered resistance, the electric current increases and enables a current to activate a sounder that generates an audible sound during the moment of sparking. But when the electric current is turned off, a problem arises in that the filings still cohere. A device was needed to loosen the filings so as to allow the next action. The Englishman Oliver Lodge developed a decoherer or trembler, which automatically dislodged the clumped filings by simply tapping on the glass tube, thus restoring the device's sensitivity and preparing for the next charge of current from the transmitter. This cumbersome device of coherer/decoherer remained in use through the nineteenth century. With the move to radio for the transmission of complex sound, the coherer was eventually replaced in the twentieth century by the vacuum tube.

The primary problem, however, in the wireless transmission of electromagnetic waves was finding a means of transmitting these waves over long distances. Hertz's original spark-gap transmitter had sent waves only a few yards. To conquer the problem of distance, Marconi (in collaboration with the British Post Office) experimented with a number of methods, including lengthening the antennas of the receiver and transmitter, setting them vertically, and connecting them to the ground. Using these innovations, in 1896, Marconi sent a Morse code message across level land and over a hill in England to a distance of three miles. In that same year, he sent the first signal over open water, across the Bristol Channel in the west of England. In 1899, he transmitted over the English Channel. In 1901, in competition with the transatlantic telegraph cable, Marconi sent a wireless message from England to Newfoundland. In 1903, from a Marconi station in Cape Cod, Massachusetts, President Theodore Roosevelt sent a wireless to King Edward VII of England.

With the pioneering work in England of establishing functional wireless communication between nations came the benefits of wireless communication for ships at sea, especially important to England as a maritime nation. In 1897, Marconi sent the first ship-to-shore wireless message. He soon established the first permanent wireless station for shipping on the Isle of Wight off the coast of southern England. Then he established a number of Marconi stations along the coast of England. Such stations were eventually set up on coastlines throughout the world. These receiving and transmitting stations meant that steam-powered ocean liners, freighters, and warships could now be constantly in touch with land as well as with each other. Steamships now had wireless operators on board to communicate emergencies and to be contacted by other ships in distress. In the most celebrated instance early in the twentieth century, the wireless operator on the sinking *Titanic* was able to send a message to the nearby ship *Carpathia*, enabling that liner to come to the rescue of at least some passengers. And by the beginning of World War I, the British Navy, like other world powers, installed wireless as essential to command at sea.

One noteworthy effect of the Marconi wireless system was that ships at sea could receive accurate time signals based on Greenwich Mean Time calculated at the Royal Observatory in Greenwich, London. Formerly dependent on the chronometer, an accurate clock carried on shipboard to ascertain longitude, ocean-going ships could now use the universal time signal of GMT transmitted globally by wireless. In this striking example of the continuity and continuing influence of British innovation, at the end of the century wireless telegraphy provided a standardized time originating in England for steamships around the globe, much as the wires of the electric telegraph had provided a standardized time for the railways of the early-Victorian age.

MODERN INVENTIONS, ENGLISH CONTRIBUTIONS

The last several decades of the nineteenth century saw the emergence of the familiar technologies of our own lives: devices for the recording and replaying of sound; and the automobile and the airplane, both powered by the internal-combustion engine. As innovation progressed globally at the end of the Victorian age, England contributed significantly to the development and the broadened use of these inventions.

Gramophone

Along with developing devices for the transmission of sound, the late nineteenth century saw a global search for machines that would record and replay sound. The major breakthrough came with the work of Thomas Alva Edison in America. Typically, one innovation generated another. Edison's recording device emerged from his work on the telephone. Looking for a way to record and then resend telephone messages, Edison hit upon using a needle controlled by the varying vibrations of voice to emboss indentations on tin foil stretched over a cylinder. Another needle following the indentations in the tin foil would play back the sound. The cylinder design was rapidly improved by the use of wax for the impression of sound vibrations. These wax cylinders were marketed as recordings of music by the celebrities of the day. There was even an effort at selling Edison talking dolls, a toy that only became successful in our own day. The new technology spread globally and was quickly adopted in England. By the 1890s, the use of wax cylinders for office dictation had become widespread. Much of the narration of Bram Stoker's *Dracula*, published in 1897, is provided through Dr. Seward's diary of events which, the text notes, is "Kept on phonograph."[6]

For all its popularity, Edison's wax cylinder had certain disadvantages, notably that the wax was not amenable to the central industrial process of mechanized reproduction. The wax cylinder was soon replaced by the mass-produced phonograph record that could be played on what the nineteenth century called the gramophone. In the 1890s, a German immigrant to the United States, Emile Berliner, developed the use of a zinc disc that could be embossed with grooves that registered sound. Through the mid-Victorian technology of electroplating, the original record could be made into a stamper that could generate innumerable exact reproductions. The general principle of replaying recorded sound by a needle moving along grooves in a disc remained in use until the advent of the computer-age compact disc (or CD) in which a laser reads microscopic variations in the surface of a disc.

England played a significant role in the spread of the new gramophone technology. In 1897, one of the world's first recording companies, the Gramophone Company, was formed in England to manufacture records and gramophone machines based on Berliner's patents. In 1898, the Gramophone Company made its first recording, a song sung by a barmaid of a London pub. The company continued to buy patent rights to more efficient methods of recording. At the end of the century, the Gramophone Company had become established as a leader in Europe in adapting the new technologies for recording sound and in the mass production of records.

Internal-Combustion Engine

In the late nineteenth century, industrial nations sought ways to replace the cumbersome steam engine dependent upon coal, the central machine of the early industrial revolution. One new source of energy was electricity obtained from coal-powered electrical generators. But with petroleum now easily obtainable through underground drilling, world attention focused on developing a compact, self-contained internal-combustion engine powered by derivatives of oil. Late-Victorian British inventors contributed to the creation of this engine, but their achievements have not been fully recognized.

In an internal-combustion engine, fuel is burned within the engine itself, rather than in an external boiler, as in the steam engine. Just as the steam engine transforms the expansive power of heated water into mechanical energy, so the internal-combustion engine transforms the explosive chemical energy of gasoline into mechanical energy. In the internal-combustion engine, the combustible gas mixture within a cylinder is ignited, most commonly with spark plugs. The explosive combustion of fuel pushes a piston within the cylinder. The piston's movement turns a crankshaft that then turns car wheels or an airplane propeller via a chain or a drive shaft. In the 1870s, the German inventors Nikolaus Otto, Gottlieb Daimler, and Karl Benz developed an efficient four-stroke gasoline-powered internal-combustion engine. The principal English contribution to this international effort was that of British engineer James Atkinson, who in 1882 invented the Atkinson cycle engine that employed a complex crankshaft to increase engine efficiency.

The gasoline-powered internal-combustion engine in which the gasoline mixture is introduced into a cylinder then ignited by a spark is, however, rather inefficient in transforming the chemical energy of gasoline into the mechanical energy of the pistons. In the late nineteenth century a more efficient form of the internal-combustion engine, the compression-ignition engine, came into being. In this device, air and relatively inexpensive heavy oil are introduced into the cylinder, then compressed. No spark is needed, as the compression of the air and oil mixture causes an explosion of the fuel that moves the piston.

The compression-ignition engine is usually called the diesel engine, after the German engineer Rudolph Diesel who did experimental work on this form of engine in the 1890s. But given the difficulty in assessing precedence in the globalized culture of innovation of the late nineteenth century, the history of technology has failed to recognize that the compression-ignition engine should instead be called the Ackroyd Stuart engine since it was first developed by the English engineer Herbert Ackroyd Stuart several years before Diesel's work. In 1886, Ackroyd Stuart built a prototype compression-ignition oil engine and patented the device in that same year. In 1891, he leased the patent rights to the English engineering firm of Richard Hornsby and Sons. This firm then built the first compression-ignition engine. In the next year, this new form of machine was set to practical work in a waterworks. It was not until 1893 that Rudolph Diesel, who had been working on a similar engine, took out a German patent.

In the last Victorian decade, pioneering work continued in England in developing the internal-compression engine invented by Ackroyd Stuart. The Hornsby firm manufactured tractors powered by this engine, and continued to sell internal-compression engines well into the twentieth century. In particular, British engineers maintained the Victorian engineering tradition of improving rail transportation. Employing the Ackroyd Stuart compression-ignition engine, in 1896 the engineering firm now called Ruston & Hornsby built the world's first oil-engined railway locomotive in England. This locomotive, able to generate nine and one-half horsepower, was used successfully within the Woolwich Arsenal, England's main producer of armaments. Ruston & Hornsby continued to build these locomotives powered by compression-ignition engines until 1902.

Automobile

The development of the compact and efficient internal-combustion engine and the compression-ignition engine made possible the emergence at the turn of the nineteenth century of the two dominant forms of modern transportation, the automobile and the airplane. British engineers contributed to this development, but as the century came to a close the site of invention had shifted from England to the European continent, especially Germany, and to the United States.

In the early-Victorian period, the British had revolutionized transportation by creating the steam-powered railway system. Mid-century England even saw practical steam-powered land vehicles, such as the steam tractors used to pull farm machinery displayed at the Great Exhibition (see Chapter 3, "Machinery as Spectacle"). But by the end of the century, as the age of steam faded, there was an intense effort within the industrial nations to apply the technology of the internal-combustion engine both in spark-ignition and compression-ignition form to self-propelled personal land transportation. In the evolution of the automobile, England took a lesser role in relation to the German inventors who survive in the names of contemporary automobile companies. In 1885, Gottlieb Daimler of Germany developed what came to be the modern gasoline automobile engine with vertical cylinders and gasoline injected through a carburetor. One year later, he built the first four-wheeled gas-powered vehicle. In 1886, Karl Benz patented a gas-powered automobile.

Like the internal-combustion engine itself, the automobile is the result of innovations made throughout the industrial world. Indeed, this process continues as the shape and efficiency of the automobile evolves within our own global car culture. At the end of the nineteenth century, the main British contributor to automobile technology was Frederick William Lanchester. Lanchester's most influential invention was the automobile disc brake, which he patented in the 1890s. In the disc brake, a disc, usually made of cast iron or ceramic composites, is connected to the automobile wheel. To stop the wheel, friction material in the form of pads is forced against both sides of the disc. Friction causes the disc and attached wheel to slow or stop. Although Lanchester used these disc brakes in the cars produced in his Birmingham factory, it took another half century for his innovation to be widely adopted and become the primary form of automobile braking.

Applying his own innovations in braking and in the construction of the internal-combustion engine, in 1895, Lanchester built the first four-wheel gasoline-driven automobiles produced in Britain. In 1899, Lanchester and his brothers formed the Lanchester Engine Company in Birmingham to manufacture cars of his own design. The company remained in business into the twentieth century, producing both automobiles for the domestic market and armored cars for the British Army in World War I.

Airplane

The late nineteenth century also saw the international pursuit of heavier-than-air human flight, the goal of imitating the birds that goes back to the Greek myth of Icarus and Daedalus. Lighter-than-air flight with balloons had been achieved in the eighteenth century, but a machine that could fly remained elusive. Self-propelled heavier-than-air flight demands three elements: an air frame that will provide lift (the flow of air under a fixed wing that provides the upward movement of the airplane), moveable controls, and a self-contained means of propulsion. In 1903, the American Wright brothers were able to apply to flight the innovations in the internal-combustion engine, as Daimler, Benz, and Lanchester had to land transportation. The Wright brothers employed the newly invented internal-combustion engine to provide propulsion. They used an airframe designed to provide lift through curved wing surfaces, as well as a tail structure designed to provide control of flight. Yet the structure of the airplane that left the ground at Kitty Hawk owed much to the Wright brothers' careful study of the theory and practical work of an early-Victorian British inventor, George Cayley.

As an amateur experimenter in the late eighteenth and early nineteenth century, Cayley, a wealthy British aristocrat, provided both the theoretical basis of aeronautics, the science of flight, as well as a practical demonstration of non-powered flight or gliding. Without any formal scientific training (the physics of flight had not yet been developed), Cayley worked out the physical dynamics of flying. In 1809 and 1810, Cayley published three papers in which he quite correctly pointed out that lift is generated by a region of low pressure on the upper surface of the wing and that curved wing surfaces generate lift more efficiently than a flat surface. In particular, Cayley was able to calculate through experimentation with a number of models the mathematics of lift provided by a fixed wing. He thus ascertained the area of wing surface necessary for sustained flight.

Cayley also worked out the essential configuration of what became the modern airplane. He saw that to provide moveable controls, it was necessary to set a vertical stabilizer at the tail with a moveable steering rudder to control sideways motion and a horizontal stabilizer with flaps or elevators to regulate the pitch or up-and-down motion. He constantly experimented with the shape of the fixed wing so as to maximize lift and advocated the

Cayley glider. A 1973 reconstruction of George Cayley's man-carrying glider. This heavier-than-air flying machine is similar to that used in 1853 to fly Cayley's coachman 900 feet over a small valley in the first recorded flight of a person in an aircraft. [Permission of Science Museum/Science & Society Picture Library.]

setting of one wing over another in the biplane frame that the Wright brothers would eventually employ. It was this theoretical and practical work that the Wright brothers drew upon in designing their airframe. Cayley realized that true flight could not be achieved until a lightweight engine was developed to give the thrust or forward motion that is necessary to generate lift. His prediction was fulfilled in the airplane that following his structural principles and employing the new internal-combustion engine flew at Kitty Hawk.

A typical Victorian inventor-as-tinkerer, in 1853, fifty years before the Wright brothers flew at Kitty Hawk, Cayley constructed a flying machine that put his theories into practice. Cayley built his machine of wood. It had fixed wings and both vertical and horizontal moveable tail controls. Since there was no self-contained propulsion, this airframe resembled what we would now call a hang glider; flight was initiated by jumping from a height. As an aristocrat, Cayley had a certain degree of control over his retainers and so, as the story goes, he directed his coachman to lie on the top of this proto-airplane for its flight. Fortunately for the coachman, the flight was successful. After initiating flight by leaping from an elevation, the coachman was carried by the glider 900 feet over a small valley. This was the first recorded flight of a person in an aircraft. Cayley's mid-Victorian machine can be seen as the first practical heavier-than-air flying machine. Although Cayley's glider has been reproduced for flight in our own time, true self-propelled flight had to wait for the Wright brothers in the first years of the twentieth century.

CONCLUSION: GAIN AND LOSS

The rise of the machine in Victorian England was truly transformational. By 1850, England had become the first fully industrialized nation and the model for the mechanized world that we still inhabit. Although nineteenth-century industrialization had certain continuities with the eighteenth century, as in the development of the steam engine and of automatic textile machinery, in the 1830s and 1840s, innovation and entrepreneurship had come together in the integrated factory system. The Victorians brought into being the processes of industrialization that we now take for granted: the replacement of animal and muscle power by steam and, by the end of the century, electricity; mechanized mass production; ceaselessly accelerating technological innovation. From the burst of inventive energy in nineteenth-century England there emerged the devices that shape the twenty-first century: the computer, the internal-combustion engine, the wireless, the gramophone, the automobile, and the airplane.

As we look back on this extraordinary narrative of invention and innovation in Victorian England, several questions remain. How can we account for emergence of industrialization at a specific moment on a small island off the coast of Europe? In this transformation to a mechanized world, what has been gained and what has been lost?

HOW ENGLAND CAME TO RULE THE WORLD

There are a number of theories that attempt to explain why the triumph of the machine took place in England in the nineteenth century. Perhaps there is no single cause, but a constellation of causes—material, cultural, and even religious—that accounts for the rise of the machine at this time and place.

For one, the island held the natural resources essential to industrialization. Situated in a temperate and wet climate, England was endowed with rivers that flowed year round and thus could provide the water power for the first textile mills. Under the Midlands, South Wales, and the Scottish lowlands were large deposits of coal, iron ore, and tin that were the raw materials necessary for industrialization. These coal and iron deposits could be easily transported by rivers, canals, and eventually the railways to the factories centralized in the Midlands. Furthermore, there was already in place an active if small-scale mining industry that could be readily expanded as the need for these minerals increased.

But raw materials are just that, raw. The transformation of coal and iron into steam engines and productive factories needs human effort, and human effort must draw upon a sustaining financial and political environment.

In nineteenth-century England, as in our own time, technological innovation depended upon financial capital to fund invention and its applications. The expansion of England's empire in the eighteenth century and the nation's profitable overseas commerce that included the slave trade provided profits that were available at the beginning of the nineteenth century for investment in domestic business such as the building of mills and the expansion of railways. At the start of the nineteenth century, too, improvements in agricultural methods had created a surplus of laborers in the countryside who were eager for work in the new industrial cities. England's overseas trading network ensured ready-made contacts for the importing of raw materials such as cotton and the exporting of industrially produced goods to markets around the world. The supremacy of England's navy after the final victory in 1815 in the wars with France assured that British shipping could freely trade around the world.

At the beginning of the nineteenth century, England was a politically stable nation with power vested not in a king but in Parliament. In the Victorian period, this parliamentary power expanded as the right to vote was gradually extended to finally include males of the working class. Although there were in the Victorian age political protests by the working class, such as the Chartist movement of the 1830s and 1840s and some industrial strikes, the violent warfare between workers and capitalists that disrupted continental nations was avoided. Furthermore, in contrast to continental nations and to the United States, England experienced no wars on its soil through the nineteenth century. Victorian England's wars were mostly fought in the territories of the empire to maintain its markets and its supply of raw materials. Since the decline of the Luddite movement of the 1810s in which technologically unemployed workers destroyed new textile machinery and burned factories, mill owners knew their property was safe. Such domestic stability under the rule of law was an essential precondition of industrial development.

The Victorians forged these favorable elements—natural resources, financial capital, an abundant labor supply, globalized trading connections, political stability—into an industrial nation in a burst of inventive and entrepreneurial energy that drew its strength from England's political, cultural, and religious traditions. England's success was due not only to financial capital, but also to intellectual capital.

Unlike continental nations such as France, England did not have a system of strong centralized authority based in the capital city. Rather than in government-sponsored academies or think tanks, English invention developed as a matter of individual initiative operating without government support. This scientific tradition also emphasized hands-on experimenting over theorizing. And it prized the challenge to conventional authority generated by individuals working alone or with a small team. In what is termed the scientific revolution of the seventeenth century, British experimenters such as Robert Boyle, often called the founder of modern chemistry, developed an effective vacuum pump and proved that sound is carried by air. Working alone, Isaac Newton formulated the laws of universal gravity and motion.

In the late eighteenth century and early nineteenth century this investigative curiosity of the experimenter was fused with the profit-seeking zeal of the entrepreneur to create the industrial revolution. Richard Arkwright developed new textile machinery such as the water frame for spinning and then combined this machine with other advanced machinery in his highly profitable textile mills. James Watt turned his experiments with the steam engine into the large and highly profitable firm of Boulton & Watt that through the nineteenth century manufactured advanced steam engines for the nation and the world. A British scientist such as Michael Faraday exemplified this tradition of fusing innovation and commerce. In the 1830s, he made small devices in his workshop to demonstrate that electric current could be generated through creating an electromagnetic field. He then worked assiduously to transform the small-scale experiments of the laboratory into machinery for large-scale, commercial electrical generation.

The culture of invention in nineteenth-century England that so strongly valued the fusion of individual effort, useful scientific applications, and entrepreneurial zeal is in large part due to the religious culture of the nation. Ever since Henry VIII broke with the Roman Catholic Church in the mid-sixteenth century, England has been an intensely Protestant country. Protestantism in the form adopted by the English incorporates a belief system, generally termed the Protestant work ethic. Rather than centering the practice of religion on retreat from the material world, the Protestant work ethic sees activity in the world, such as the improvement of textile machinery or the building of textile mills, as a vocation infused with religious meaning.

All people are called by God to their duty in life. Pursuing vigorously the duty or occupation that one finds oneself in is to follow the divine plan and thus success in one's work in the world is a sign of spiritual grace. The riches of a businessman or the prosperity of a mill owner are signs that he is carrying out God's Will.

In Victorian England, where evangelicalism was the dominant religious force, industrialists and inventors alike drew energy from the belief that they were carrying out the Will of God. Furthermore, an ethic that claimed work as a moral duty helped to maintain a dutiful and compliant work force within the factories.

Although grounded in Protestant religious belief, the Victorians' ethical valuing of commercial and technological accomplishment was expressed in secular form. The term that captured the Victorian version of the Protestant work ethic was the word, still in use, "self-help." The expression comes from an enormously popular book published by Samuel Smiles in 1859, *Self-Help, with illustrations of Conduct and Perseverance*. The sale of a quarter million copies through the end of the nineteenth century attests to the attractiveness of an ideal that celebrates British industrial supremacy, assumes this primacy is due to the British tradition of a moralized work ethic, believes that individual effort can enable inventors to rise from the lowest social ranks, and assumes that machine technology emerges from the craft tradition. Smiles is worth quoting at some length as he expresses the cultural grounding of the Victorian rise of the machine:

> One of the most strongly-marked features of the English people is their spirit of industry, standing out prominent and distinct in their past history, and as strikingly characteristic of them now as at any former period. It is this spirit, displayed by the commons of England, which has laid the foundations and built up the industrial greatness of the empire. This vigorous growth of the nation has been mainly the result of the free energy of individuals, and it has been contingent upon the number of hands and minds from time to time actively employed within it, whether as cultivators of the soil, producers of articles of utility, contrivers of tools and machines, writers of books, and creators of works of art.... [T]he duty of work is written on the thews and muscles of the limbs, the mechanism of the hand, the nerves and lobes of the brain.[1]

For Smiles, the exemplars of self-help in England were the technologists. His very Victorian celebration of the inventor and entrepreneur as heroes of the age is exemplified in the section of *Self-Help* devoted to "Leaders of Industry—Inventors and Producers." Here, he singles out "the laborious, patient, never-tiring James Watt, the mathematical-instrument maker."[2]

For Smiles, Watt exemplifies the practical genius and the ethic of work in developing "the king of machines"[3]: "Watt was one of the most industrious of men; and the story of his life proves, what all experience confirms, that it is not the man of the greatest natural vigour and capacity who achieves the highest results, but he who employs his powers with the greatest industry and the most carefully disciplined skill."[4] Smiles went on to write a number of equally popular works including a *Life of George Stephenson* (1857), the man who built the Rocket steam locomotive that ran on the Liverpool and Manchester Railway in 1830, and a more general work on Victorian mechanists, a multi-volume *Lives of the Engineers* (1861–62).

If the values of a work ethic and individual initiative encouraged individual invention and entrepreneurship for the Victorians, the central government did play its part in encouraging and protecting financial investment. The construction of large mills and the expansion of the railway system required the raising of substantial sums of money. In the early-Victorian period there emerged the financial institution of the joint stock company in which investors bought a specific amount of stock in a developing company. The Joint Stock Companies Act of 1844 made it easier to found such companies and thus to raise capital. But investment in joint companies was hampered because investors carried unlimited liability, that is, investors were personally responsible for any debts incurred by the stock company beyond the amount personally invested in stock. To eliminate potential financial disaster for the individual stockholder and to encourage investing, Parliament in 1855 passed the Limited Liability Act legislating that a shareholder in a limited company is not personally liable for any of the debts of the company beyond the value of his investment in that company. This limited liability investment system has proved highly efficient in raising what we now call venture capital and is still in use.

LOSS AND GAIN: THE AGE OF MACHINERY IN EVERY OUTWARD AND INWARD SENSE OF THAT WORD

Just as in our time an older generation can recall the world before e-mail and the cell phone, so the Victorian generation that came to maturity in the 1830s and 1840s could remember the world before railways, factories, and inexpensive machine-made furniture and clothing. And in these first decades of industrialization, amid the sudden rise of industrial cities and the cycles of poverty created by overproduction, the Victorians debated whether the age of the machine that had so suddenly come into being was really superior to the pre-industrial England of agricultural villages and hand craft. And from the perspective of our age of electricity and computers, we too can consider, to use a popular Victorian phrase, the loss and gain resulting from the nineteenth-century rise of the machine.

The Victorians were supremely aware that they lived in a world radically transformed by machine technology. To return to the words of the influential social critic Thomas Carlyle in his aptly titled essay, "Signs of the Times," in 1840: "Were we required to characterise this age of ours by any single epithet, we should be tempted to call it ... above all others, the Mechanical Age. It is the Age of Machinery in every outward and inward sense of that word."[5] Carlyle's attention to both the "outward and inward" epitomizes the sense of the Victorians that not only their "outward" material life, but also their "inward" psychological life had been transformed by living in an age dominated by the machine.

Certainly, material life in the nineteenth century improved dramatically. For the Victorians, the word "Progress" (often used with a capital "P") summed up their belief in the inevitable continuation of technological change and that such change could only be beneficial. Thomas Babington Macaulay, a prominent Victorian cultural critic and historian writing in 1830, saw the poverty of the industrial towns as only a blip in the inevitable historical movement toward improvement in the lives of all. In a Victorian science-fictional mode, he imagines an England one hundred years in the future. He envisions that

in the year 1930 a population of fifty millions, better fed, clad and lodged than the English of our time, will cover these islands ... that cultivation, rich as that of a flower garden, will be carried up to the very tops of Ben Nevis and Helvellyn [British mountains], that machines constructed on principles yet undiscovered will be in every house, that there will be no highways but railroads, no traveling but by steam ... that men would be in the habit of sailing without wind, and be beginning to ride without horses.[6]

Macaulay's prophesy was, if anything, too cautious. The English were sailing the oceans without wind and riding the roads in horseless carriages even before the end of the Victorian age.

Certainly, Macaulay was correct in pointing to the positive changes brought by technological innovation, particularly as government regulations modified the harsh conditions of factory labor. By mid-century, the railway had replaced the stagecoach traveling over muddy roads. Rural villages were no longer isolated. Even the urban working class could take day excursions by train. Mechanization had made the nineteenth-century English better fed, the result of increased agricultural production using steam-powered tractors and threshing machines. Fresh fish could be shipped inland by rail. Mutton was more readily available even to the working class as meat was quickly transported from the imperial dominions of Australia and New Zealand in refrigerated steamships.

At home, fresh food such as milk also became widely available, delivered to the cities quickly and without spoilage by the railway system. The radius of London's "milk shed" had gradually expanded so that by the end of the century the rail system had reached out two hundred miles to dairy farmers. Urban dwellers now spoke of "railway milk." Thomas Hardy's novel *Tess of the d'Urbervilles* (1891) includes a striking scene from a rural perspective depicting the outreach of the industrial urban world to the countryside. Tess, now a milkmaid, travels by horse-cart to deliver the milk cans to the regular train arriving to pick up the fresh milk to take to the city: "A fitful white streak of steam at intervals upon the dark-green background denoted intermittent moments of contact between their secluded world and modern life. Modern life stretched out its steam feeler to this point three or four times a day, touched its native existences, and then quickly withdrew its feeler again."[7]

There is no argument that private lives were improved in many ways. The mechanization of textile production brought down the cost of cotton and woolen clothing so that wool stockings and cotton dresses were now available to all classes. Although the housing of the industrial workers who poured into the unprepared manufacturing towns of Manchester and Birmingham was at first horrific, the gradual regulation of housing by government and the rising wages of manufacturing jobs created a nation of urban and suburban homeowners by the end of the century. These homes were well-furnished with the inexpensive machine-carved wooden furniture and the electroplated knives and forks similar to those shown at the Great Exhibition.

The conditions of labor were mixed through the century, but generally grew more salutary as the century drew on, particularly as government regulation moderated the harshness of early-Victorian factory work. There were periods of unemployment through the Victorian period due to the effects of business cycles, overproduction by increasingly efficient mechanical methods, and unforeseen foreign events such as the American Civil War. But these times of distress were temporary and the trend of job creation was upward. Technological employment early in the century had impoverished craft workers, as in the example of the rural hand-loom weavers. But as factory production in the great manufacturing towns increased, those whose crafts had been eradicated by mechanization found work as employment grew in the new industrial economy.

In general economic terms, a new class system was generated by the rise of the machine and the consequent centralization of mechanized production in urban factories and mills. By its very nature, the factory system created a new class of urban industrial labor to operate the factories and a new class of mill owners who provided the large sums of capital for these industries. The economic results of the system were mixed and are still being debated.

As government over time allowed workers to unionize, the dynamic of power between workers and owners became more balanced. Through struggles, including strikes and negotiations between capital and labor, wages rose and employment became more stable. With increased government regulation generated by pressure from labor, working hours were limited for children, women, and men. Compulsory safety controls reduced industrial accidents and improved the air quality both within and beyond the factories. But for all these improvements, the class system created by industrial capitalism generated enormous disparities of wealth, with the owners taking the profits of production and the workers who did not share in these profits limited to a set wage.

With industrialization, employment opportunities for women expanded. The early textile mills drew in women who although working long hours in unsafe conditions were able to add to the family income. In the later Victorian period, the commercial applications of the telephone and of the newly invented typewriter opened up office work for women as telephone operators and what the age called "typewriters," what we would call secretaries. Such respectable employment in offices rather than factories contributed to the emergence of the liberated or, in late-Victorian terms, the "New Woman," whose foundation was financial independence.

If we look to what Carlyle called the "inward" effects or the consequences for consciousness of the rise of the machine the historical verdict is also mixed. Certainly, with mechanization the very nature of work changed. And the psychological and moral effects of this transformation of labor became a major source of debate among the Victorians.

For the Victorian industrialists and their apologists, of whom Andrew Ure was paramount, the mechanization of labor was seen in ethical terms as freeing productive work from human frailty. Machines seemed to obey the Protestant work ethic more diligently than workers. In *The Philosophy of Manufactures*, Ure praises "the progression of the British system of industry, according to which every process peculiarly nice, and therefore liable to injury from the ignorance and waywardness of workmen, is withdrawn from handicraft control, and placed under the guidance of self-acting machinery."[8] For James Nasmyth, famous in his time for inventing the steam hammer, automatic devices provided mechanical alternatives to human weakness: "Machine tools were found to be of much greater advantage [over workmen]. They displaced hand-dexterity and muscular force; they were unfailing in their action; they could not possibly go wrong in planing and turning, because they were regulated by perfect self-acting arrangements; they were always ready for work, and never required a Saint Monday [not coming to work on Monday]."[9] For the mill owners, the deskilling that resulted from the division of labor became a virtue in creating stability both within the factory and the society. Like most Victorians, Ure was constantly fearful of a revolution by the industrial workers and argued that the more skillful the worker becomes the "more self-willed and intractable he is apt to become, and, of course, the less fit a component of a mechanical system, in which, by occasional irregularities, he may do great damage to the whole."[10] Here the "system" refers both to the smooth operation of factory production as well as the smooth running of the social system that makes industrial production possible.

For the defenders of mechanization, the imposition of factory discipline was justified as a heroic moral effort aimed at replacing the work habits of agricultural laborers with a new work ethic. "To devise and administer a successful code of factory discipline, suited to the necessities of factory diligence, was the Herculean enterprise, the noble achievement of Arkwright," who established the first mechanized cotton mill.[11] Industrialism was to these Victorians an ethical enterprise in its "training human beings to renounce their desultory habits of work, and to identify themselves with the unvarying regularity of the complex automaton."[12] And yet Ure also recognized the intractable nature of the human mind in resisting factory discipline. In speaking of reshaping the minds of those who had moved from country village to industrial town, from hand looms to power looms, he lamented: "It is found nearly impossible to convert persons past the age of puberty, whether drawn from rural or from handicraft occupations, into useful hands."[13]

With the reconceptualizing of the human body as an engine that burns fuel to produce energy, Victorian debates on the value of mechanized labor centered on the issue of fatigue, the using up of bodily and mental energy. It was recognized that routinized work can drain the human battery of bodily and mental energy by running against the innate human desire for individuated action. On the other hand, it was argued by defenders of industrialization that the machine-body connection could be harmonized so as to reduce the expenditure of energy. Indeed, Ure argued that mechanized labor by its nature was less fatiguing than pre-industrial work such as that of the hand-loom weaver: "In the factory, every member of the loom is so adjusted, that the driving force leaves the attendant nearly nothing at all to do, certainly no muscular fatigue to sustain.... Whereas the non-factory weaver, having everything to execute by muscular exertion, finds the labour irksome, makes in consequence innumerable short pauses, separately of little account, but great when added together."[14]

Apologists of industry even extended their belief in the energizing effects of mechanization to children working in the cotton mills. For them, the mill becomes an industrial utopia tuning children's muscular power to their work. One observer saw them "taking pleasure in the light play of their muscles." Equally important is that the pace provides some moments during work in the mill for the children to refresh their energy; the visitor delighted "to see them at leisure, after a few seconds' exercise of their tiny fingers, to [see them] amuse themselves in any attitude they chose, till the stretch and winding-on [of yarn on spools] were once more completed ... As to exhaustion by the day's work, they evinced no trace of it on emerging from the mill in the evening, for they immediately began to skip about."[15]

For the defenders of factory discipline such as Ure and Charles Babbage, inventor of the Victorian proto-computer, any mental and bodily fatigue in the factory would fade over time as human consciousness was transformed by familiarity with the machine. Optimistically, they imagined that the mental engine of the human brain could be improved, much as the steam engine had been improved, through innovation. They prophesied, perhaps incorrectly, that the capacity for enduring the fatigue of routinized work increases as workers became habituated to factory discipline. Babbage writes that in the cases of "mental exertion; the attention bestowed on the new subject not being so perfect at first as it becomes after some exercise."[16] They prophesied that as society become industrialized the worker would become more accustomed to repetitive work and indeed create a new form of expertise through this familiarity: "The constant repetition of the same process necessarily produces in the workman a degree of excellence and rapidity in his particular department, which is never possessed by a person who is obliged to execute many different processes."[17]

For other Victorians of the industrial age, it seemed clear that mechanized work expands production, but degrades the psyche. According to these latter-day Luddites, who continued the opposition to labor-saving machinery, the worker was being transformed into a human machine.

In *Hard Times*, his industrial novel, Charles Dickens vividly describes in his imagined Coketown the regularized, mechanized life lived according to the factory clock: "It [Coketown] contained several large streets all very like one another, and many small streets still more like one another, inhabited by people equally like one another, who all went in and out at the same hour, with the same sound upon the same pavements, to do the same work, and to whom every day was the same as yesterday and to-morrow, and every year the counterpart of the last and the next."[18]

To these critics of mechanization, the machine worker is seen as tending an unvarying master in an unchanging routine and making a standardized product from standardized material. With the use of power looms, especially the programmed jacquard loom, the process of weaving becomes automatic as the shuttle carries the woof threads relentlessly and tirelessly through the warp. There is no place for personal aesthetic choice. The attention to irregular detail in the material and the judgments for eye and hand demanded by craft are lost. The worker has to adapt to the regular, unceasing, and unchanging rhythm of the power loom.

Furthermore, as factory labor was divided in specialized and repetitive activities, the worker was removed or alienated from the finished product such as the completed textile. The mill-worker only runs a carding machine or tends the power loom rather than seeing the creation through from raw material to finished shirt. The pride the craftsman takes in the object he or she has completed is gone. The Victorians called factory workers "hands," a term that nicely encapsulates the transformation of the body in factory labor so that only a fragment of the body was needed. The unified energy of the body and certainly of the creative mind at play in the labor of the hand-loom weaver was no longer necessary.

Some contemporary commentators were aghast at what they saw as this dehumanization of work in the mills. Karl Marx, in *Capital: A Critique of Political Economy* (1867), saw mechanized labor as wholly destructive of the worker's mind and body:

> In the factory we have a lifeless mechanism independent of the work-man, who becomes its mere living appendage. The miserable routine of endless drudgery and toil in which the same mechanical process is gone through over and over again, is like the labour of Sisyphus. The burden of labour, like the rock, keeps ever falling back on the worn-out labourer. At the same time that factory work exhausts the nervous system to the uttermost, it does away with the many-sided play of the muscles, and confiscates every atom of freedom, both in bodily and in-tellectual activity.[19]

This sense of damage to the inward life caused by tedious repetition of a limited task found its best-known nineteenth-century expression in the writings of John Ruskin, the most influential of the Victorian anti-machine voices. In a parody of Adam Smith's influential praise of the division of labor in the work of the pin-maker, Ruskin pictures a glass-bead maker as manifesting the diseased madness generated by repetitive action. Here Ruskin calls attention to the psychic sacrifices demanded of the worker in satisfying the Victorian lust for commodities:

> Glass beads are utterly unnecessary, and there is no design or thought employed in their manufacture. They are formed by first drawing out the glass into rods; these rods are chopped up into fragments of the size of beads by the human hand, and the fragments are then rounded in the furnace. The men who chop up the rods sit at their work all day, their hands vibrating with a perpetual and exquisitely timed palsy, and the beads dropping beneath their vibration like hail.[20]

Yet, what critics of the machine such as Ruskin failed to appreciate is that psychically destructive, routinized work such as cutting glass rods into beads could be taken over by machinery, thereby freeing the worker from such enervating activity. Despite its tedium, for the generations that moved from farm to city, the indoor work of the mill was less exhausting and debilitating than digging and plowing in the rain in the endless drudgery of agricultural labor.

For all the Victorian nostalgia for hand craft seen in such critics as Ruskin, the rise of the machine did not destroy, but only relocated the site of craft and of pride in work. In the Victorian age, as in our time, people took pleasure in their skilled work with machines. Operating the power loom was demanding work, and the men (women were excluded from this assignment) who did such work were proud and were paid well. The locomotive driver piloting his train through a storm, the steam-hammer operator shaping iron into form, and the steam-crane operator lifting a bridge section into place took delight in their expertise in operating complex machinery, much as operators of heavy machinery such as construction cranes do in our own machine age.

Furthermore, with the use of such devices as the jacquard loom where production follows a programmed path, the question arises of where creativity is located as manufacturing becomes automated. In hand-loom weaving, the individual weaver integrated both designing and the work of setting the pattern into cloth, often incorporating small design changes as the work of weaving progressed. In the mechanized process, the worker as machine watcher does not participate in the creation of the patterns to be woven. Yet, the activity of human creation is not lost, but relocated to the designer who sketches the pattern to be woven. This design is then translated into a pattern of holes in punched cards that regulate the loom. Removing this split between designer and workman built into automation became a chief goal of the major Victorian reactions against mechanization, the Arts and Crafts movement of the later nineteenth century that sought to resurrect handicraft methods of production. Nor did anti-machine critics appreciate the fusion of intellect and imagination involved in designing the machines that performed with consistent accuracy.

Associating beauty only with hand work, such as hand-carved furniture and hand-loomed rugs that are irregular and rough, fails to register the new shape of beauty in the smooth and regular lines of a metal pitcher or vase shaped by the machine in multiple identical forms. To limit one's sense of beauty to the rough-hewn gargoyles carved by medieval craftsmen on a cathedral façade, as John Ruskin did in his influential essay, "The Nature of Gothic," is to exclude the grace in the regular glass panels and iron columns of the Crystal Palace or the great train sheds of the London railway terminals. To look back with nostalgia on the world before steam power is to fail to see the energy of the steam age exemplified in Turner's evocation of the sublime beauty and intense force of the locomotive racing over the landscape in his *Rain, Steam, and Speed.*

As technological innovation continues with increasing velocity into our own day, we, like the Victorians, worry about the tyranny of mechanization.[21] The debate between Luddites and advocates of the machine continues. Public discussion in the computer age asks many of the same questions that the Victorians asked in the age of steam. Is the mind essentially a machine, in our time a computer? Can freedom of the will survive as the machine becomes dominant? Is there a place for creative work? Will imagination be lost to the logic of mechanism? Is there a place for simple daydreaming? Is a human and humane life still possible with the rise of the machine?

NOTES

INTRODUCTION

1. Somerville, *Autobiography*, 235–36.
2. Marx, Leo, "Technology," 978.
3. Carlyle, "Signs of the Times," 27:59.

CHAPTER 1

1. Babbage, *Economy of Machinery*, 21.
2. Ure, *Philosophy of Manufactures*, 344.
3. Ibid., 26.
4. Marx, *The Machine in the Garden*.
5. William Wordsworth, "On the Projected Kendal and Windermere Railway," ll. 1–5.
6. William Wordsworth, "Steamboats, Viaducts, and Railways," ll. 9–10.
7. Dickens, *Dombey and Son*, 120.
8. The painting can be seen in the online collection display of London's National Gallery at http://www.nationalgallery.org.uk/collection/artist/default.htm.
9. Tennyson, "Locksley Hall," ll. 181–82.
10. Mumford, *Technics and Civilization*.
11. "Edmund Cartwright," http://www.cottontimes.co.uk/cartwrighto.htm.
12. Cunningham, *Industrial Revolution*, 632.
13. Cardwell, *History of Technology*, 186.
14. Ure, *Philosophy of Manufactures*, 153.
15. Ibid., 159.
16. Baines, *History of the Cotton Manufacture*. In Freedgood, *Factory Production*, 71.

17. Karl Marx, *Capital*, 113.
18. Nasmyth, *Autobiography*, 162.
19. Dickens, *American Notes*.
20. Smith, *Wealth of Nations*, 85.
21. Mumford, *Technics and Civilization*.

CHAPTER 2

1. Dorothy Wordsworth, *Recollections*, 121.
2. Ure, *Philosophy of Manufactures*, in Freedgood, *Factory Production*, 153.
3. Spufford, "The Difference Engine," 270.
4. Hyman, 49.
5. Babbage, *Economy of Machinery*, 261.
6. Ibid., 138.
7. Ibid.
8. Swade, *Difference Engine*, 83.
9. Ibid., 82.
10. Ibid., 99–100.
11. Babbage, *Passages*, 30.
12. Swade, *Difference Engine*, 85.
13. Swade, *Charles Babbage*, 23.
14. Gibson and Sterling, *Difference Engine*, 83.
15. Babbage, *Passages*, 102.
16. Swade, *Difference Engine*, 105.
17. Swade, *Charles Babbage*, 31.
18. Babbage, *Passages*, 88.
19. Cardwell, *History of Technology*, 188.
20. Lovelace, "Sketch of the Analytical Engine," 122.
21. Ibid., 100.
22. Swade, *Difference Engine*, 166.
23. Lovelace, "Sketch of the Analytical Engine," 122.
24. Lardner, "On the Application of Machinery." 264.
25. Coalbrooke, "Address," 57.
26. Lovelace, "Sketch of the Analytical Engine," 122.
27. Babbage, *Passages*, 100.
28. Ibid., 90–91.
29. Ure, *Dictionary of Arts*, 367.
30. Ibid., 83.
31. Ibid., 362–63.
32. Ibid., 362–63.
33. Ibid., 376.
34. Ibid., 358.
35. Ibid., 367.
36. Gaskell, *Artisans and Machinery*, 24.
37. Sussman and Joseph, "Prefiguring the Posthuman" and Hayles, *How We Became Posthuman*.

38. Ure, *Encyclopedia of Manufactures*, 13.
39. Nasmyth, *Autobiography*, 242.
40. Rabinbach, *Human Motor*, 1–2.
41. Ibid., 2.
42. Cardwell, *History of Technology*, 166.
43. Baines, *History of the Cotton Manufacture*, 243.
44. Babbage, *Encyclopedia of Manufactures*, 23.
45. Gaskell, *Artisans and Machinery*, 165.
46. Ibid., 218.
47. Ibid., 220.

CHAPTER 3

1. Williams, *Keywords*, 165–68.
2. Tennyson, "Locksley Hall," l. 184.
3. Gibbs-Smith, *The Great Exhibition of 1851*, 19.
4. Ibid., 16–17.
5. Ibid., 17.
6. Ibid., 31.
7. Ibid., 83.
8. Ibid., 82.
9. Ibid., 83.
10. Ibid., 80.
11. Ibid.
12. Auerbach, *The Great Exhibition of 1851*, 104.
13. Altick, *The Shows of London*, 72–76.
14. Mayhew, *The World's Show. 1851*, 160–61.
15. Freedgood, *Ideas in Things*, 147.
16. Williams, *Keywords*, 78–79.
17. *The Crystal Palace Exhibition*, 63.
18. Ibid.
19. Gibbs-Smith, *The Great Exhibition of 1851*, 90.
20. *The Crystal Palace Exhibition*, 69.
21. Gibbs-Smith, *The Great Exhibition of 1851*, 19.
22. *The Crystal Palace Exhibition*, 201.
23. Ibid., 193.
24. Ibid.
25. Ibid., 65.
26. Gibbs-Smith, *The Great Exhibition of 1851*, 26.
27. Richards, *Commodity Culture*, 31.

CHAPTER 4

1. Standage, *Victorian Internet*, 50.
2. Ibid., 51.
3. Somerville, *Autobiography*, 236.

4. Warburg, *Industrial Muse*, 59–60.

5. Menke, 90.

6. Standage, *Victorian Internet*, vii.

7. Ibid., 160.

8. Ibid., 80–81.

9. North River Steamboat.

10. The painting can be seen in the online collection display of London's National Gallery at http://www.nationalgallery.org.uk/collection/artist/default.htm.

11. Kipling, "Deep-Sea Cables," 173.

12. Kipling, "McAndrew's Hymn," 120–26.

13. Cotton Famine Road.

14. Portsmouth Royal Dockyard.

CHAPTER 5

1. Engels, *Condition of the Working Class*, 50.

2. Ibid., 61.

3. Ibid., 60.

4. Pritchard, 155.

5. Life of the Industrial Worker.

6. Ibid.

7. Carlyle, *Past and Present*, 11–12.

8. The Luddites.

9. Ibid.

10. Life of the Industrial Worker.

11. Macaulay, "Review of Southey's *Colloquies*," 1702.

12. Development Trusts Association.

13. Ibid.

CHAPTER 6

1. Shelley, *Frankenstein*, 171–72.

2. Thames Tunnel.

3. Dickens, *Dombey and Son*, 120.

4. UK Telephone History.

5. Maxwell.

6. Stoker, *Dracula*, 93.

CONCLUSION: GAIN AND LOSS

1. Smiles, *Self-Help*, 17.

2. Ibid., 19.

3. Ibid., 18.

4. Ibid., 19.

5. Carlyle, "Signs of the Times," 27:59.

6. Macaulay, "Review of Southey's *Colloquies*," 1701.

7. Hardy, *Tess of the D'Urbervilles*, 187.
8. Ure, *Philosophy of Manufactures*, viii.
9. Nasmyth, *Autobiography*, 309.
10. Ure, *Philosophy of Manufactures*, 20.
11. Ibid., 15.
12. Ibid.
13. Ibid.
14. Ibid., 7.
15. Ibid., 301.
16. Babbage, *Economy of Machinery*, 122.
17. Ibid., 123.
18. Dickens, *Hard Times*, 12.
19. Marx, *Capital*, 117.
20. Ruskin, "The Nature of Gothic," 185.
21. Giedion, *Mechanization Takes Command*.

GLOSSARY

Analytical Engine: proto-computer designed by Charles Babbage but never built.

Androides: machine resembling a human being.

Arc light: device in which light is generated by a spark that arcs or jumps between two electrically charged carbon elements.

Automatic machinery: devices that operate without the need for human intervention, usually on the feedback principle.

Cast iron: metal created by fusion of iron ore and carbon.

Chartism: movement of workers in late 1830s that presented to Parliament a petition or charter asking for increased parliamentary representation.

Coal gas: gas created by burning coal; used for lighting.

Coke: residue of coal after heating.

Coke-fired blast furnace: furnace fueled by coke; used to make cast iron.

Combination: union or organization of workers.

Commodity: an object for sale.

Compression-ignition engine: power source in which increasing pressure on a mix of heavy oil and air in a cylinder causes an explosion that moves the piston; often called diesel engine.

Cyberpunk: form of science fiction focusing on advanced information technology and marginal people.

Cyborg: entity fusing the mechanical and the organic.

Difference Engine: proto-computer for calculation designed by Charles Babbage but never fully built.

Dynamo: large-scale device generating electricity in wire coils rotating within a magnetic field.

Electric telegraph: communication system that transmits an electric current over wires from location to location; interruptions in current create signals that can be translated into the dots and dashes of Morse code.

Electromagnet: iron in which electric current produces magnetism.

Electromagnetism: describes relation between electricity and magnetism, particularly that a changing magnetic field produces an electric current and that a changing electric field produces magnetism.

Electroplating: a process by which a thin coating of metal is applied to the surface of another metal by the action of electricity.

Factory: large workshop incorporating mechanized processes; powered by water wheel, steam engine, or electricity.

Faraday disc: experimental device built by Michael Faraday. Two wires connected to a copper disc; when rotated through the poles of a horseshoe magnet an electric current is generated in the wires connected to the disc.

Feedback: The return of a portion of the output of a process or system to the input; used to maintain performance or to control a system or process, as in Watt's steam-engine governor.

Fluff: small particles of cotton waste floating in the air of textile mills; caused respiratory illness.

Functionalism: architecture theory holding that a building should be designed primarily to carry out the functions or needs of that building rather than to display ornament.

Galvanism: early term for electricity.

Gatling gun: hand-cranked machine gun.

Gauge: distance between tracks of a railway.

Ghost in the machine: theory the human body is controlled by a soul that is not material and thus not subject to physical law.

Gramophone: system for sound reproduction using grooved disc and needle.

Greenwich Mean Time (GMT): standard time set at Greenwich Royal Observatory.

Gunboat: iron warship of shallow draft used on inland waters.

Hand-loom weavers: craftsmen who worked in their homes on hand-powered looms.

Incandescent lighting: lighting device using electric current running through a filament within a vacuum in a glass bulb.

Induction: the producing or inducing of electricity in a conductor, usually a wire circuit, as a result of a changing magnetic field about the conductor.

Industrial capitalism: economic system in which control of industrial facilities such as factories rests with the owners rather than with the workers or the government.

Industrial novels: novels generally written in the 1840s about the conditions of life and the relations between the classes in the manufacturing towns.

Internal-combustion engine: power sources in which a spark explodes a mix of gasoline and air in a cylinder; explosion then moves a piston.

Ironclad: armor-plated, steam-powered warship.

Jacquard loom: weaving device directed by a program punched into paper cards.

Laissez faire: theory that government should not interfere with the natural laws of economics.

Limited-liability joint stock company: business entity in which the financial responsibility of the contributor is limited to the amount of the investment used to buy stock.

Luddism: movement by workers in early nineteenth century to destroy new textile machines that were replacing hand work.

Luddite: modern term for person opposing new technology.

Machine gun: automatic rapid-fire weapon.

Maxim gun: machine gun with water-cooled jacket.

Mechanism: theory that organic bodies are subject to the same laws of physics that govern physical objects.

Mechanist: an inventor or experimenter; also defender of mechanization.

Mule: machine invented in late eighteenth century for more efficient spinning of yarn.

Paddle-wheel: steam-driven wheel containing paddles or oars mounted at the stern or sides of a ship for propulsion.

Papier-mâché: material made from paper paste mixed with glue that can be molded into various shapes.

Paternalism: benevolent control by owner of factories and workers' communities.

Power loom: water- or steam-powered device for weaving.

Pulley block: wooden block holding a metal wheel with a groove in which a rope can run; used for ropes controlling sails.

Pumping engine: steam engine and pump combined for raising water; often used to keep mines dry.

Railway time: standardized national time used by railways; eventually based on Greenwich Mean Time.

Safety valve: device automatically controlling pressure within boiler of steam engine.

Screw propeller: power-driven shaft with radiating blades placed underwater at stern of steamship to provide propulsion.

Self-acting: functioning without need for human intervention.

Shuttle: draws threads through warp in weaving.

Sleeper: ties holding together tracks of the railway.

Spinning jenny: multi-spool spinning wheel invented in late eighteenth century that enabled a worker to operate many spools at one time.

Stationary engine: steam engine set in one location for such uses as powering a factory or pumping water from a mine.

Steamboat: steam-powered vessel used on inland waterways.

Steam engine: device that supplies power through transforming the expanding steam created by boiling water into mechanical energy, generally by driving a piston within a cylinder.

Steam-engine governor: automatically regulates entry of steam from boiler to engine.

Steam hammer: large steam-driven mallet, used to beat metal into shapes or drive piles into the ground.

Steamship: ocean-going vessel usually of iron propelled by steam.

Telephone: electrically-powered system used to transmit and receive sound by variations in the electrical current.

Thermodynamics: science of the relations between heat and other forms of energy.

Turbine: machine in which the flow of a fluid over blades imparts a rotary motion as in a water wheel or dynamo.

Underground railway: trains running in tunnels beneath urban areas; first powered by steam, then by electricity.

Undersea cable: protected wires laid underwater to carry telegraph or telephone signals.

Voltaic pile: early battery of metal discs separated by paper.

Water frame: machine for spinning yarn invented in late eighteenth century that substituted water power for human power in operating frames for spinning yarn.

Water wheel: device that uses flowing or falling water to create power by paddles mounted around a wheel mounted outside a factory; power is transmitted to machinery via the shaft of the wheel.

Wireless: system sending electromagnetic waves through the air; as in electric telegraph, interruptions in current create signals that can be translated into the dots and dashes of Morse code; also called wireless telegraph.

BIBLIOGRAPHY

Altick, Richard D. *The Shows of London*. Cambridge, MA: Harvard University Press, 1978.

Auerbach, Jeffrey A. *The Great Exhibition of 1851: A Nation on Display*. New Haven, CT: Yale University Press, 1999. Good scholarly account of the making of the Great Exhibition.

Babbage, Charles. *The Economy of Machinery and Manufactures*. New York: New York University Press, 1989.

———. *Encyclopedia of Manufactures*. In *Works of Charles Babbage*.

———. *Passages from the Life of a Philosopher*. New Brunswick, NJ: Rutgers University Press, 1994.

———. *Works of Charles Babbage*, edited by Martin Campbell-Kelly. 11 vols. London: Pickering, 1989.

Baines, Edward. *History of the Cotton Manufacture in Great Britain. : With a Notice of Its Early History in the East, and in All the Quarters of the Globe; a Description of the Great Mechanical Inventions, which Have Caused Its Unexampled Extension in Britain; and a View of the Present State of the Manufacture, and the Condition of the Classes Engaged in Its Several Departments*. [1835]. New York: Augustus M. Kelly, 1966. Also in Freedgood, *Factory Production*.

Baum, Joan. *The Calculating Passion of Ada Byron*. Hamden, CT: Archon, 1986. Useful biography of Ada Lovelace.

Berg, Maxine. *The Machinery Question and the Making of Political Economy, 1815–1848*. Cambridge: Cambridge University Press, 1980. Standard account of the relation of mechanization to political change.

Cardwell, Donald. *The Norton History of Technology*. New York: W. W. Norton, 1994. Excellent concise history of technology from earliest times.

Carlyle, Thomas. *Past and Present*. Boston: Houghton Mifflin, 1965.

———. "Signs of the Times." In *Works of Thomas Carlyle,* edited by H. D. Traill, vol. 27. New York: AMS Press, 1979.

Channell, David F. *The Vital Machine: A Study of Technology and Organic Life.* New York: Oxford University Press, 1991.

Coalbrooke, Henry Thomas. "Address on Presenting the Gold Medal of the Astronomical Society to Charles Babbage." In *Works of Charles Babbage,* vol. 2.

Conrad, Joseph. *The Secret Agent.* New York: Penguin, 2007.

"Cotton Famine Road." http://www.dingquarry.co.uk/location—geography/ cotton-famine-road.asp.

The Crystal Palace Exhibition: Illustrated Catalogue, London 1851. An Unabridged Republication of the Art-Journal Special Issue. New York: Dover, 1970.

Cunningham, William. *The Industrial Revolution.* Cambridge: Cambridge University Press, 1908.

Development Trusts Association. http://www.dta.org.uk.

Dickens, Charles. *Dombey and Son.* London: Penguin, 1985.

———. *Hard Times.* New York: W. W. Norton, 1990.

"Edmund Cartwright and the Power Loom." http:www.cottontimes.co.uk/ cartwrighto.htm.

Engels, Friedrich. *The Condition of the Working Class in England.* Translated by W. O. Henderson and W. H. Chaloner. Stanford: Stanford University Press, 1963.

Freedgood, Elaine, ed. *Factory Production in Nineteenth-Century Britain.* New York: Oxford University Press, 2003. Very useful compilation of nineteenth-century documents about mechanized production.

———. *The Ideas in Things: Fugitive Meaning in the Victorian Novel.* Chicago: University of Chicago Press, 2006.

Freeman, Michael. *Railways and the Victorian Imagination.* New Haven, CT: Yale University Press, 1999. Inclusive account of the effect of the railway on Victorian thought and art.

Gaskell, Peter. *Artisans and Machinery: The Moral and Physical Condition of the Manufacturing Population Considered with Reference to Mechanical Substitutes for Human Labour.* London: John Parker, 1836. Reprint London: Frank Cass, 1968.

Gibbs-Smith, C. H. *The Great Exhibition of 1851: A Commemorative Album.* London: His Majesty's Stationery Office, 1950.

Gibson, William, and Bruce Sterling. *The Difference Engine.* New York: Bantam, 1991.

Giedion, Sigfried. *Mechanization Takes Command.* New York: Oxford University Press, 1948. Classic account of the widespread effects of mechanization.

Hardy, Thomas. *Tess of the D'Urbervilles.* New York: Signet, 1999.

Hayles, N. Catherine. *How We Became Posthuman: Virtual Bodies in Cybernetics, Literature, and Informatics.* Chicago: University of Chicago Press, 1999. Influential critical discussion of new ways of thinking in the current age of the cyborg.

Hyman, Anthony. *Charles Babbage: Pioneer of the Computer.* Princeton, NJ: Princeton University Press, 1982. Definitive biography of Babbage.

Jennings, Humphrey. *Pandæmonium 1660–1886: The Coming of the Machine as Seen by Contemporary Observers*. Edited by Mary-Lou Jennings and Charles Madge. London: Papermac, 1995. Fine collection of nineteenth-century responses to the rise of the machine.

Kipling, Rudyard. *Rudyard Kipling's Verse: Definitive Edition*. Garden City, NY: Doubleday, 1940.

Lardner, Dionysus. "On the Application of Machinery to the Calculation of Astronomical and Mathematical Tables. Charles Babbage." *Edinburgh Review* 59 (July 1834): 263–327.

The Life of the Industrial Worker in Nineteenth-Century England. Victorian Web. http://www.victorianweb.org/history/.

Lovelace, Ada. "Sketch of the Analytical Engine invented by Charles Babbage Esq. By L. F. Menabrea, of Turin, officer of the Military Engineers, with notes upon the memoir by the translator." In *Works of Charles Babbage*, vol. 3.

The Luddites. http://www.spartacus.schoolnet.co.uk/PRluddites.htm.

Macaulay, Thomas Babington. "Review of Southey's *Colloquies*," in *The Norton Anthology of English Literature*. Edited by M. H. Abrams and Stephen Greenblatt. Seventh ed. New York: W. W. Norton, 2000.

Marx, Karl. *Capital: A Critique of Political Economy*. In Freedgood, *Factory Production*.

Marx, Leo. *The Machine in the Garden: Technology and the Pastoral Ideal in America*. New York: Oxford University Press, 1964. Classic account of the response to the machine in nineteenth-century America.

———. "Technology: The Emergence of a Hazardous Concept." *Social Research* 64 (Fall 1997): 965–88.

Maxwell, James Clerk. http://heritage.scotsman.com/jamesclerkmaxwell/.

Mayhew, Henry, and Cruickshank, George. *The World's Show. 1851: or, The Adventures of Mr and Mrs Sandboys and Family, Who Came Up to London to "Enjoy Themselves" and to See the Great Exhibition*. London: George Newbold, 1851.

Menke, Richard. *Telegraphic Realism: Victorian Fiction and Other Information Systems*. Stanford: Stanford University Press, 2008.

Mumford, Lewis. *Technics and Civilization*. New York: Harcourt, 1963. Classic account of the centrality of the clock to the rise of industrialism.

Nasmyth, James. *James Nasmyth, Engineer: An Autobiography*. Edited by Samuel Smiles. New York: Harper & Brothers, 1884. Reprint Bradley, IL: Lindsay, 1989.

North River Steamboat. http://en.wikipedia.org/wiki/Clermont_(steamboat).

Portsmouth Royal Dockyard Historical Trust. http://www.portsmouthdockyard.org.uk.

Pritchard, R. E. *Dickens's England: Life in Victorian Times*. Phoenix Mill, UK: Sutton Publishing, 2002.

Rabinbach, Anson. *The Human Motor: Energy, Fatigue, and the Origins of Modernity*. Berkeley: University of California Press, 1992. The definitive account of the idea of the human body as engine and its social implications.

Richards, Thomas. *The Commodity Culture of Victorian England: Advertising and Spectacle 1851–1914*. Stanford: Stanford University Press, 1990.

Ruskin, John. "The Nature of Gothic." In *The Genius of John Ruskin: Selections from His Writings*, edited by John D. Rosenberg. Charlottesville: University of Virginia Press, 1997.

Shelley, Mary. *Frankenstein*. New York: W. W. Norton, 1996.

Simmons, Jack. *The Victorian Railway*. London: Thames and Hudson, 1995. Comprehensive account of the building and the effects of the railway system.

Smiles, Samuel. *Life of George Stephenson and of His Son Robert Stephenson*. New York: Harper & Bros., 1868.

———. *Lives of the Engineers*. London: J. Murray, 1874.

———. *Self-Help, with illustrations of Conduct and Perseverance* [1859]. London: IEA Health and Welfare Unit, 1997.

Smith, Adam. *The Wealth of Nations*. In Freedgood, *Factory Production*.

Somerville, Alexander. *The Autobiography of a Working Man*. In Jennings, *Pandæmonium*.

Spufford, Francis. "The Difference Engine and *The Difference Engine*." In Spufford, *Cultural Babbage*.

Spufford, Francis, and Uglow, Jenny, eds. *Cultural Babbage: Technology, Time and Invention*. London: Faber and Faber, 1996. Fine collection of essays on Victorian technology.

Standage, Tom. *The Victorian Internet: The Remarkable Story of the Telegraph and the Nineteenth Century's On-line Pioneers*. New York: Walker, 1998. Fascinating account of the building of the national and international communications systems in the nineteenth century.

Stoker, Bram. *Dracula*. New York: Broadview, 1998.

Sussman, Herbert. *Victorians and the Machine: The Literary Response to Technology*. Cambridge: Harvard University Press, 1968. General study of Victorian literature dealing with the machine.

Sussman, Herbert, and Gerhard Joseph. "Prefiguring the Posthuman: Dickens and Prosthesis." *Victorian Literature and Culture* 32 (2004): 617–28.

Swade, Doron. *Charles Babbage and His Calculating Engines*. London: Science Museum, 1991. Fascinating account of the failed effort to build the Difference Engine in the nineteenth century and its successful construction in the twentieth century.

———. *The Difference Engine: Charles Babbage and the Quest to Build the First Computer*. New York: Viking, 2001. Good general account of Babbage's work on his proto-computers.

Tennyson, Alfred Lord. *Poems of Tennyson*, edited by Christopher Ricks. Berkeley: University of California Press, 1987.

Thames Tunnel. http://en.wikipedia.org/wiki/Thames_Tunnel.

Turner, Joseph Mallord. http://www.nationalgallery.org.uk/collection/artist/default.htm.

UK Telephone History. http://web.ukonline.co.uk/freshwater/histuk.htm.

Ure, Andrew. *The Philosophy of Manufactures: or, An Exposition of the Scientific, Moral, and Commercial Economy of the Factory System of Great Britain*. New York: A. M. Kelley, 1967. And in Freedgood, *Factory Production*.

————. *A Dictionary of Arts, Manufactures, and Mines; Containing a Clear Exposition of Their Principles and Practice*. New York: D. Appleton, 1844.

Warburg, Jeremy. *The Industrial Muse: The Industrial Revolution in English Poetry*. London: Oxford University Press, 1958. Useful collection of poems dealing with industrialization.

Williams, Raymond. *Culture and Society, 1780–1950*. New York: Columbia University Press, 1983.

————. *Keywords: A Vocabulary of Culture and Society, Revised Edition*. New York: Oxford University Press, 1985. Useful in tracing how the meaning of important and commonly used words changed with industrialization.

Woodward, Llewellyn. *The Age of Reform: England 1815–1870*. Oxford: Oxford University Press, 1962. Fine general history of early and mid-Victorian England.

Wordsworth, Dorothy. *Recollections of a Tour Made In Scotland*. In Jennings, *Pandæmonium*.

Wordsworth, William. *The Poems*. New Haven, CT: Yale University Press, 1981.

INDEX

Page numbers for illustrations are in boldface.

About the Author

HERBERT SUSSMAN has published extensively in the area of Victorian literature and culture. His works include *Victorians and the Machine: The Literary Response to Technology* and most recently *Victorian Masculinities*. He has taught at the University of California, Berkeley, and Northeastern University. Currently he is an adjunct professor at the New School.